HUMAN ECOLOGY OF SIKKIM

HUMAN ECOLOGY OF SIKKIM

A Case Study of Upper Rangit Basin

Brigadier Kuldip Singh Gulia

KALPAZ PUBLICATIONS

Delhi-110052

Human Ecology of Sikkim
A Case Study of Upper Rangit Basin

Rs. 850

ISBN: 81-7835-325-3

Published in 2005 in India by
Kalpaz Publications,
C-30, Satyawati Nagar, Delhi-110052
E-mail : Kalpaz@hotmail.com

Laser Typesetting at : Shiva Infosystem, Delhi
Printed at : Mehra Offset Press, Delhi

Contents

LIST OF MAPS TABLES AND FIGURES

Preface

Sikkim is a land of pristine beauty with unique culture caused by the diffusion of Buddhism, Hinduism and Animistic rituals of different religious faiths. It is still relatively untouched by the burgeoning human development activities as is seen in the rest of India. The Upper Rangit Basin is the spiritual Centre in the Himalayas in general and Sikkim in particular through the destiny of religious centres. The residents of this secluded habitat, which on account of its isolation, have continued to live and function in a traditional life style in accordance with their respective ethnic and religious customs. It thus affords ideal holistic view of all human ecological aspects resulting from the interrelationship between humans, the biotic life and the abiotic physical environment.

The objective of this research was to study and analyze all facets constituting the human ecological aspects of Upper Rangit Basin, an important mountain enclave in the Eastern Himalayas in the Indian state of Sikkim.

Religious faith among the local folk has been the mainstay, and living beacon in their day-to-day life, festivities and rituals, even at the cost of personal and state development projects. The lamas and monks reign supreme, and the simple ethnic folk consider the lama's word as 'law'. However, the people of the Basin, especially the youth have the keen desire to modernize; but this is at the cost of losing out on their traditional knowledge of ecology and customs gathered over generations of learning and centuries of experience and symbiotic interrelationship. A mutual incompatibility and antagonisms between traditions and modernity leads to societal conflict.

This conflict can have wide ramifications on development and policy paradigms of the state government. This research has been focused on arriving at solutions based on human ecology of the various human and geographic ecosystems in Sikkim, i.e. Deep Ecology, Cultural Ecology and Spiritual Ecology. This could be the basis for the intellectuals, government officials and religious heads plan development activities based on a healthy inter-mix of traditional ecological knowledge with modern scientific techniques.

Acknowledgements

I gratefully express my gratitude to Dr. Rangadhar Sahu who effectively motivated me to exert the efforts necessary to compile, analyze and to maintain a balanced work flow in the writing of this thesis. I am also grateful to the director, faculty members and the documentation staff at the Centre for Himalayan studies (NBU) who collectively sparked enthusiasm, and showed genuine interest in my work.

I am deeply indebted to Dr. Maitreyi Choudhury who sitmulated innovative methodologies in field work and data collection, I express my deepest gratitude to Dr.Jyoti Prakash Tamang and Mrs Namrata Tamang of Sikkim Government College for giving insight in the field of ethnobotany. Dr.Ajay Rastogi and Dr. Ruchi Pant were a great motivating factor in selection of study and research material and I acknowledge the same with deepest gratitude. Iam grateful to Dr. Eklavya Sharma and his team of experts at G. B. Pant Institute (Sikkim Unit) for providing insight into all environemnt related aspects in the area of work.. I am thankful to Dr. Jyotsana Subba, Deputy Director, Bureau of Economics and Statistics, Government of Sikkim for providing me necessary data and information. I express my gratitude to Mr.S.D Lama for his help in recognition of plant species in the area of Yuksom.

My thanks are due to the lamas and monks at all the monasteries in the Upper Rangit Basin for introducing me to their respective spiritual traditions, wherein Captain Yongda of Pemayangtse monastery contributed voluntarily with a great fervour. I convey my sincere thanks to Mr. Punnet Kansal, IAS, ex-SDM Ravangla, and later ADC West Sikkim in facilitating my area survey, and to his wife Amrita, for her assistance in survey of Ravngla market.

I express my deepest gratitude to Inspector Genral G.S Uban, VSM and Brigadier K.N Mehra of the Special Frontier Force for their permission and constant encouragement in completion of this thesis.

My special thanks to all officers, leaders and men of Project Archer, who generated positive attitudes during the progress of my work.

Lieutenant Colonel Mani Ram Pankaj took on additional work responsibilities, sparing me time for research and study. Dafadar Clerk S.K Mishra assisted by Paljor untiringly typed the drafts and the final script. Tashi and Kunga assisted in ground and monastic survey and in preparation of charts and maps. Sangye, my driver negotiated all the mountain roads in helping me to reach out to all possible places in the study area. Passing and Tsering provided me with hot beverages at most untimely hours by night and day. I express my deepest gratitude to each one of them.

Throughout the writing of this thesis the Sikh Regiment especially 4 SIKH and 6 SIKH were at the center of my hear. I am grateful to the Sikh Regiment for making me worthy of taking on to writing and research related projects.

I would like to thank my wife Santhosh, who was a force multiplier in keeping my enthusiasm at peak. To facilitate my work, she continued to bear the pangs of separation, and did so without a grumble. My daughter Dipti was of a great assistance in listing out bibliographical details from my collection of books and papers. My other daughters Kirti and Taru organized my work table to bring in orderliness. My sons-in-law Captain Lokesh and Lieutenant Subhash(Indian Navy) kept my morale high by cooperating with their respective wives. My grandson Rijul (Teetoo) brought in freshness with his frequent smiles.

Lastly I wish to convey my gratitude to my respected father, brothers, and their families for their constant help and unflinching support. My mother who left for heavenly abode showered spiritual enthusiasm from above during the entire period, and she continues to give shade of the spiritual umbrella in the form of a *tathagata*.

1
Introduction

SECTION - I
THE PROBLEM

Principle, Concept and Philosophy

The physical human environment, in a broader sense "the biosphere" has been of an absolute importance for the quality of life of human societies. Throughout the evolutory history of the mankind, it has been only recently that human development began to threaten its very existence. Does the humankind have an absolute right over nature? This has, of recent, cropped up as one of the fundamental issues of environmental ethics, and has been reflected in some of the environmental principles agreed upon by the members of the United Nations Organisation (UNO) (Glaeser 1995).[1]

Principle 1 of the Stockholm Declaration on the Human Environment (UNCHE : 19) states, "Man has the fundamental right to freedom, equality and adequate conditions of life, in an environment of a quality that permits a life of dignity and well being, and he bears a solemn responsibility to protect and improve the environment for present and future generations—". This statement highlights both, the fundamental human right, and also the implied human responsibility. Principle 4 supports this interpretation of Principle 1, "Man has a special responsibility to safeguard and wisely manage the heritage of wild life and its habitat. Nature conservation must therefore receive importance in planning for economic development".

Ecology and (economic) development have been clubbed together to form the concept of "ecodevelopment". Deriving the purpose of this clubbing, we may conclude that development is not solely to enhance economic growth, it must be sustainable, and should preserve and improve the environmental basis for economic development. This is therefore clear from the UN Conference on the Human Environment (UNCHE) and the Stockholm Declaration of 1972 that Development Policy has to go hand in hand with nature and environment. Thus their recommendation is "Development policy in accordance with nature and environment".

The 1992 United Nations Conference on Environment and Development (UNCED) acknowledged the fact that humanity's relationship to nature and the use of resources and ecological services, have seldom been considered important. The scientific discipline of human ecology provides a framework for thought and research along these lines from political, theoretical, ethical, ethnic, cultural and habitat perspectives.

Reflecting on above, there thus is a need of a philosophy that has a radical departure from the current Western trends. The humankind needs an eco-philosophy that strengthens the nature's roots in the Earth. The grid to be penetrated now, more than at earlier turning points, is one that has grown dangerously abstract in relation to human needs. The urgent task now is to regain concreteness. The concrete world is a value saturated creative process of Oriental Buddhists and Hindus, and Western philosophers, central to the development of eco-philosophy. (Kvaloy 1994).[2]

Environmental Influence and Impact on Environment

Even though technology has removed us from close touch with nature and reduced a certain amount of environmental hazards, our physical surroundings can still exert an influence. Some natural hazards are actually intensified by modern culture. Our modern society has chosen to build large scale multipurpose dams on rivers astride major seismic earthquake zones in developing and developed countries, producing catastrophe of much greater proportion than a conservative folk culture will face in that area in centuries of time period. (Jordan 1990).[3]

Modern culture makes some heavy demands on environment. This is true even in the realm of recreation and tourism. Since early fifties leisure time and related recreational activities by certain rich plainsmen have increased greatly in the Himalayan states/countries. The demand for "wilderness" recreational zones has risen sharply in the last quarter century, and no end to the increase seems at hand.

Hikers, campers, hunters, fishers, bikers, snowmobile enthusiasts, weekend cottagers, mountain climbers, boaters, sightseers and others are making unprecedented demands on the Himalayan states. Persons engaged in wildlife trade, and those engaged in illegal felling of trees and other fauna have also been drawn in. Such massive activities have put a stop on conservation of the physical environment, let alone nurture it. Mountain roads now suffer from traffic jams, and the mountain towns suffer from residential congestion, litter and noise pollution. The study of Darjeeling hill areas (which forms the southern fringe of Rangit Basin catchment areas) revealed substantial environmental damage including soil loss and long term soil deterioration. Erosion of soil during monsoons, and water scarcity during lean seasons has become a common feature. In less congested wilderness, such as is available in the Upper Rangit Basin, hordes of hikers can beat down trails to the extent that vegetation is altered, erosion encouraged, and wildlife diminished. Even the best intentioned, conservation-minded visitors do some damage.

Ethical Dimension

The starting point for an environmental ethical system is a critique of the instrumental relations between humans and the environment. In contrast to the anthropocentric ethical systems that have been so common throughout the history of philosophy, environmental ethics also makes nature the object of human responsibility. This orientation touches human ecology, which is biocentric at its core. What both have in common is the quest for a social dimension. In the philosophical tradition, the ego was enthroned in knowledge and action. The 'domination-of-nature' ideology arose out of the dissociation of cognition and action from ethical principles, and in the wake of the sheer success of science and technology. (Glaeser 1995).[4]

Environmental ethics of survival demands that humankind have to mend their ways, change their behaviour and attitudes of life, cut their material demands, record their priorities, simplify their life-styles, forego consumerist culture and harmonise with nature. Ecological ethics demand that mankind has to learn to :-

(a) Live with less; do more with less and be content with less.
(b) Produce few consumers; consume judiciously according to need, cut out greed and conserve for future.
(c) Minimise wastage in daily life, avoid criminal wastage of food, water and electricity and all other forms of energy, and
(d) 'Reduce', ' reutilize', and 'recycle' all kinds of wastes within ecologically possible limit. (Sinha 1995).[5]

The philosophical war on the subject of environmental ethics - the ethical divide between the East and West continues unabated. The strange combination of Western 'mind' in its self proclaimed instrumental and positive form, and the Eastern 'heart', which is more conservative and spiritual in nature, does not resolve gospel of a peaceful co-existence. Sharing a common platform for sustainability remains a far cry. (Bookchin 1996).[6]

Mismatch of Politics and Human Ecology

Political will for environmental protection has been conspicuous by its absence. Environmental problems are by their very nature holistic, needing collective or concerted action, which the governments find hard to deal with through normal political and diplomatic practices. Environmental issues are so different from developmental issues, and do not affect day-to-day life as the issues of poverty and economics. Their impact is hard and of very longterm to point out. Hence politics/ international politics and environmental problems fit badly together because of different nature. (Ramakant 1995).[7]

The mismatch of politics and ecological problems is portrayed below :-

(a) Politics on which the national governments are formed are based on narrowly focused short-term national/political interests, which in turn determine their worldview. The governments do disregard areas that adversely affect national interests, and stress more on their respective political party interests. By contrast, environmental problems are global, and affect nations indirectly in a longrun.

(b) In a way behavioural orientation in the study of politics has become quite dominant. This orientation leaves out values and morals that cannot be measured or expressed in economic terms. Most of the environmental problems are moral issues by their very nature.

(c) Activities of the private sector are mainly devoted to profit. In modern times the so-called economic issues, are related to scarce resources, and the private sector brings tremendous pressures on the governments to give them a free hand or greater freedom in exploring these resources.

(d) While modern politics is supposed to deal with "realism" and real subjects, ecological problems are not regarded as real or serious issues by political leaders (Ramakant 1995).[8]

The basic problem for environmental ethics, and later on, for political environmental policies is the determination of whether environmental protection should exist for the sake of humankind or whether it should exist for its own sake. One essential difference between ethical and political blue prints of earlier times and those of today lies in the fact that, in the past, the contents of human actions were known. The ethical approach that aims at the individual could be confident that healthy, mature persons could know, and had to know what was good and what was evil. If we observe the effects of human intervention into the ecosystem, we can see that this is no longer the case. The ecosystem of Planet Earth is not entirely known, and cannot be controlled instrumentally. Although we are generally aware of the hazards of pollution, for instance, even ecosystem researchers are not able to predict with any degree of certainty what the impacts of individual interactions might be. Thus in many cases the harmful effects of interventions into nature are difficult, if not impossible to estimate. It is precisely because knowledge of the effects of technology is lacking - so goes the new argumentation - that there is the ethical demand to handle nature as carefully as possible, and to intervene in the natural order as little as possible . (Glaeser 1995). [9]

The starting point for an environmental ethical system is a critique of the instrumental relations between humans and the environment. In contrast to the anthropocentric ethical systems that have been so common throughout the history of philosophy, environmental ethics makes nature the object of human philosophy. Environmental ethics makes nature the object of human responsibility. This orientation touches human ecology, which is bio-centric at its core. What both have in common is the quest for a social dimension. In the philosophical tradition, the ego was enthroned in knowledge and action (ethics). The domination of nature ideology arose out of the dissociation of cognition and action from ethical principles, and in the wake of the sheer success of science and technology. (Glaeser 1995) .[10]

Historic - Religious Dimensions

Gradually throughout the rise of agricultural civilisations during the last 10,000 years, man has increased his confidence and ability to control the means of subsistence. Religion became more a means of establishing favoured relationship with the unknown, and less a means of survival per se. Landscape became property and the concept of proprietary ownership developed. The Judaic-Christian philosophy in particular portrayed land as the God-given property of man. Nature existed for man to be used by him, and exploited for his own good. Man no longer considered himself merely one part of biotic community, interacting with it in a humble and mutually beneficial way. He was now its owner and master, capable of treating it like any other property. Christianity rose as "the most anthropocentric religion the world has seen". The egocentricity of man could be seen, however, long before the time of Christ. For example, the famous passage from Genesis 1 : 28 exemplifies favoured status which religion bestowed on man:

"And God blessed them, and God said to them, 'be fruitful and multiply, and fill the earth and subdue it, and have dominion over the fish of the sea and over the birds of the air and over everything that moves upon the earth." (Southwick 1972).[11]

The religious traditionalists of Judaic-Christian faith however contest the view that their religion has acted as a catalyst agent towards deterioration of environment. The environmental damage occurred all along throughout broad reaches of Asia where other religions thrive. Yet it is true that advance of mankind has carried with it, the possibility of some destructive force. The Asiatic and African cultures were not particularly gentle with nature. They slashed and burned, overgrazed and stripped the forests for firewood. But their power to do damage was limited. (Toffler Alvin 1980).[12]

Summarizing of Human Ecological Problems

Ecology of Popular Cultures Vis a Vis Rural Folk Culture. Eugene. P. Odum classified ecosystems according to source and level of energy, as outlined at Table 1.1. Serial 4 of the Table represents the popular culture of cities, suburbs and industrial parks. This is aptly named the 'Fuel powered Urban Industrial System', being wealth and pollution generating system, in which fuel replaces the Sun as the chief energy source. These kinds of ecosystems are parasitic, and dependent on the natural ecosystems, and ecosystems based on rural folk culture. With time, the wealth divide has increased, bringing high pressures on the quality of life of the folk cultural region dwellers. (Odum 1975)[13] . (Jordan and Rown Tree 1990).[14]

TABLE NO 1.1
ECOSYSTEMS CLASSIFIED ACCORDING TO SOURCE AND LEVEL OF ENERGY

	Annual Energy flow (Power level) Kilo-calories per Square Meter
1. **Unsubsidized Natural Solar-powered Ecosystems** Examples: open oceans, upland forests. These systems constitute the basic life-support module for spaceship earth	10000-10000 (2000)*
2. **Naturally Subsidized Solar-powered Ecosystems** Examples : tidal estuary, some rain forests. These are the naturally productive systems of nature that not only have high life support capacity but produce excess organic matter that may be exported to other systems or stores.	10,000-40,000 (20,000)*
3. **Man Subsidized Solar - powered Ecosystems** Examples : agriculture, aquaculture. These are food and fiber-producing systems supported by auxiliary fuel or other energy supplied by man.	10,000-40,000 (20,000)*
4. **Fuel-powered Urban-industrial Systems** Examples : cities, suburbs, industrial parks. These are man's wealth-generating (also pollution -generating) systems in which fuel replaces the sun as the chief energy source. These are dependent (i.e., parasitic) on classes 1-3 for life support and for food and fuel.	100,000-3,000,000 (2,000,000)

* Numbers in parentheses are estimated round-figure averages, (1975) E.P. Odum actually little more than guesses since the earth's ecosystems have yet to be inventoried in sufficient depth to calculate averages.

Inter-Ethnic Divide. According to Robert H Jackson, an ethnic group is a collection of persons who share a racial identity, occupy an ethnic platform, recognize and value their common occupancy, and are organized and therefore have a common interest in maintaining their association. (Daniel 1996).[15] When faced with other societies and ethnicities, either on fringes or in core regions of a geographical entity, acculturation and assimilation between them does not normally take place. Conflicts give rise to ethnic warfare, leading to long drawn insurgencies. Most world regions, especially South Asian and South East Asian countries are facing major ethnic turmoils, adversely affecting human development.

Population Pressures. No one cause can be attributed to environmental impact where population is concerned, however a number of critical factors which when combined, contribute significantly. They include rapid population growth, distribution of the rapid growth (often occurring in the most biologically rich and ecologically vulnerable world areas), density, migrations and tourism. They include very high rates of resource consumption and concurrent pollution. The 'Environmental Impact = Population × Consumption × Technology' equation illustrates how population interacts with environment. Most often when (P) Population and (C) Consumption increase either independently or together, Environmental (I) Impact increases. Technology (T) can contribute both, negatively (pollution) or positively (energy efficient technology). (Khan 1998).[17]

Deforestation and Allied Degradation. Forest has been reduced world wide to approximately 55 per cent of its original cover, and the rate of deforestation is in excess of 10 million h. every year. In India, the annual rate of destruction of forest is 1.3 million ha. The total global loss of species from deforestation alone is 140 species every day. Dr TI Khan explains that deforestation is causing a number of irreversible ecological changes, because of the large-scale exposure of the natural soil systems, which leads to increased erosion and an indirect impact on water resources. The link between food production, soil and water resources is more delicate in the mountain ecosystems.

Loss of Gene pool. Andrew Gouldie (1990)[17] has warned, "Aside from nuclear war, there is probably no more serious environmental threat then the continued decay of the genetic variability of crops. Once the process has passed certain point, humanity will have permanently lost the coevolutionary race with crop pests and diseases and will no longer be able to adapt crops to climatic change. DDT has major effect on sea birds. Their eggshells become thinned to the extent that reproduction fails in fish eating birds. Then, there are indirect causes of wild life decline vehicle speed, noise and mobility have upset remote and sensitive wildlife population."

Human Agency and Geomorphology. All kinds of landforms have been rendered into a state of acute and dangerous brinkmanship of disasters. Landforms produced by excavation, by construction and dumping and accelerated sedimentation, accelerated weathering and mass movement; deliberate and non-deliberate modifications of channels and river channel changes; accelerated coastal erosions and mountain avalanches and land slides and human impacts on seismicity and volcanoes have collectively and individually rendered the planet earth unsafe for all varieties of life forms.

Climatic and Atmospheric Changes. The world climatic conditions hithertofore underwent slow and gradual changes in terms of long drawn out geological periods. Human engineering activities and industrial after effects have together contributed towards accelerated changes on the global atmospheric and allied surface conditions. Prolific increases in release of carbon dioxide and other green house gases, the ozone depleting CFCs, aerosols, air pollution, acid rains, and photochemical smogs are constantly converting the air blanket and earth's surface info a toxic hell, having an irreversible adverse effect on the earth's biodiversity.

SECTION - II
HUMAN ECOLOGY AND HUMAN ECOSYSTEMS

Introduction

Eugene P. Odum,[19] the father of modern ecology defined ecosystem development in terms of three parameters: -

(a) It is the orderly process of community changes, which are directional and therefore predictable.

(b) It results from the modification of physical environment and population structure by the community.

(c) It culminates into the establishment of an ecosystem, which is as stable as is biologically possible in the area in question.

Development of ecosystems as explained above continues to be governed by the nature unless and until there is direct or indirect human intervention. Human ecosystem is conceptually different. It is largely because the massive harvests of negative inputs from human's fuel powered and energy intensive systems modify, halt or abort this developmental course.

Ecology deals with the dynamic balance of nature, with the interdependence of living and nonliving things. Since nature also includes human beings, the science must include humanity's role in the natural world - specifically the character, form, and structure of humanity's relationship with other species and with the inorganic substrate of the biotic environment. From a critical viewpoint, ecology opens to wide purview the vast disequilibrium that has emerged from humanity's split with the natural world. One of natures very unique species, *Homosapiens*, has slowly and painstakingly developed from the natural world into a unique social world of its own. As both worlds interact with each other through highly complex phases of evolution, it has become as important to speak of 'Human Ecology', as to speak of natural ecology.

'Human Ecology', explains Amos H. Hawley[19] (1994), in his write up on 'Ecology and Human Ecology', like plant and animal ecology represents a special application of the general viewpoint to a particular class of living things. It involves both, a recognition of the fundamental unity of animate nature and an awareness that there is differentiation within that unity. Man is an organism, and as such he is dependent on the same resources, confronted with the same elementary problems, and displays in essential outline the same mode of response to life conditions as is observed in other forms of life. Thus the extension of patterns of thoughts and techniques of investigations developed in the study of collective life of lower organisms to the study of man is a logical consummation of the ecological point of view.

Hawley goes on to further emphasise that, the extraordinary degree of flexibility of human behaviour makes for a complexity and a dynamics in the human community without counterpart elsewhere in the organic world. It is this that sets man apart as an object of special enquiry and gives rise to a 'human ecology' as distinct from the general study of ecology.

Definition

The International Organisation for Human Ecology (IOHE) in Vienna (1981)[20] defines human ecology as follows : that which deals "with human life in all its physical, chemical, biotic, mental, social and cultural manifestations from an ecological point of view ; this means considering the interrelationships of a human being or several human beings and the outside world surrounding him/her or them".

The IOHE believes that human ecology should proceed from Haeckel's definition of ecology as the 'ecology of the species *Homosapiens'*. The science of the interrelations among living beings and their surrounding world manifests itself as a systematic science. Interrelations and interactions depend on the traits of those affected, on the one hand, and the characteristics of the surrounding world on the other. Human ecology is understood to be one of several special ecologies, namely that of humans. The specific difference is that, "Human beings are capable of purposive action and of understanding the functioning of systems of which they are part". (IOHE, 1981).

The Compass of Human Ecology

There were some writers during pre World War II period, who would have human ecology encompass the whole field of social science, and there were others who relegated it to the status of a mere sociological research technique. (Hawley, 1944) .

The idea of 'wholeness' or 'holistic approach' to human ecology research studies took concrete shape through the works of Klaus and Buhr (1976)[21]. The works amplify that by 'wholeness' we mean the characteristics of things or systems that display a 'qualitatively distinct and autonomous behaviour with respect to their constituent parts. Concepts such as completeness, intactness and autonomy are relevant here'. Thus it is necessary to distinguish between 'wholeness' and 'totality'. While 'wholeness' relates to "qualitative definition', 'totality' (entirety) is 'quantitative composition' of elements or parts.

Bernhard Glaeser (1995)[22] and contemporary human ecologists contribute towards the qualitative holistic approach.

The holistic view on human ecology as summarized by James N. Anderson (1995)[23], "My conception of human ecology comprises all aspects of man as a biocultural animal as they are subsumed within the structure and functioning of ecosystems, including human morphology, reproduction, population genetics, stress, physiology, nutritional requirements, the ecology of health and disease and human adaptability. It also encompasses human population ecology - that is population process; fertility, mortality migration and population structures to include: -

(a) The feedback between biological and cultural factors (technological, economic organisational) in populations.
(b) Spatial and economic distribution and ecological consequences of populations trends.
(c) Interrelationship of population's ecological perceptions, environmental cognition, language as cultural codification, ideology, spiritual inputs and their ecological applications, creative thinking, planning and decision making.
(d) Rural, urban and industrial ecosystems, environmental aesthetics, design and planning, government-citizen interactions, and environmental law and policy making.
(e) Human interactions with biota, and earth's biotic environment.

Cultural Ecology

'Cultural Ecology' is used here to denote the interactional analysis of environmental - cultural relationships, an essentially deterministic position that has developed under the leadership of Julian Steward (1995)[24] and (1968)[25]. The definitive characteristics of cultural ecological approach are:-

(a) 'Adaptation' is the major process of cultural changes.

(b) The analysis of sociocultural-environmental adaptations serves to uncover variables that explain the origin of particular cultural features or structures in similar environmental conditions.

(c) The degree and kind of functional interdependence among the parts of culture are not equal; 'core' features, those closely interrelated with subsistence activities and economic arrangements, have casual priority over 'secondary' features.

(d) The functional relationship of relevant environmental factors causes certain organisational relationships, which in turn give rise to other relevant aspects of the culture; i.e., causation in simple, linear and one-to-one. Analysis is directed to the investigation of the environment and subsistence concerns; i.e., to the 'relevant' or 'effective' environment of the culture core.

Ethnoecology

Ethnoecology is an ethnographic endeavour. It is the study of constituent ethnic groups residing closely together in a traditional human habitat. Ethnic groups in a given geographical habitat perceive the environment in accordance with their knowledge, traditions and mind sets. Bates (1960)[26] in a compelling description of 'environment' in relation to ethnic groups states: -

(a) The environmental elements as perceived, which may be called 'perceptual environment'.

(b) Elements perceived or not, that effect the organism, which Bates terms the 'effective environment'.

(c) The environmental elements, influential or not, but are detectable and inferable, is aptly termed 'total reality'.

Vayda and Rappaport (1968)[27] summarise the characteristics of ethnoecology as follows: -

(a) Its emphasis upon perceptual environment and its general lack of serious consideration of interactions between cognitive domains or with effective environment result in narrow and least interactional approaches.

(b) Its analysis is limited to intra cultural ecological relationships.

(c) Insofar as it deals with effective environment, it seeks to evaluate and predict the effects of various behavioural possibilities on the cultural micro environment.

(d) It makes assumptions of a high degree of homogeneity and stability in cultural categorisation.

Summary of Studies of Human Ecology and Human Ecosystems

The foregoing descriptions, lead us to following assumptions that merit research and studies (Andy 1965[28], 1971[29]) :-

(a) Human populations are an integral part of most, if not all present ecosystems. Ecosystems are, as they are because of man, his numbers, his varying behaviour, and his use of energy.

(b) The nature and structure of local ecosystems is everywhere relatively recent. Although most biota predate man's entry into specific ecosystems, some biota are more recent than human invaders of various niches of the ecosystems. For instance, man himself provides few niches for parasites and micro organisms. His behaviour diverts energy flows and creates artifacts, thus providing new niches.

(c) Humans, as individuals and in populations are involved in profound reticulate transactions with physical elements of the environment as well as with biotic ones. This is readily detectable in soil and watershed characteristics.

(d) Human's transactional relations with an ecosystem can be treated from analytic and synthetic standpoints, just like those of other major components of its biota. The difference between humans and other living things is relative. Human populations are more dynamic, more manipulative and more dominant than other species in ecosystems. Complex human sociocultural behaviour enables him to collapse the time span required or biological adaptations, and has permitted an unparalleled explosive adaptive radiation into virtually all the earth's biomes.

(e) Human populations and their complex sociocultural behaviours are in constant flux. It is unnecessary to determine if a specific population and its behaviour are changing, but only to determine the rate or nature of change. The rate of change is relative. The change was relatively slow during the era of gathering and hunting. Once domestication of botanical and zoological life took concrete shape, the pace of change accelerated.

(f) The humans possessive behaviour leads to collective arms aggression, leading to war, which may be local, regional or global. The ability of human's to unite against another human group has historically affected and changed cultures of both, the victors as well as the vanquished.

A holistic integrative study of any or all the human ecosystems will therefore need to encompass all assumptions and ecosystems facets listed above.

SECTION - III
HUMAN INTERACTIONS

Interaction Oriented Models of Human Ecology

Various conceptual models have been devised that attempt to incorporate interactions taking place between human groups and the environment. A model devised by Philips -Howard (1985)[30] is depicted at Figure 1.1. The model is useful for studying the local human interactions with the environment. The model also emphasises the intrinsic interactions within the human community, and those within the local non-human ecosystems.

Figure 1.1 represents a model of interactions within human settlements. This model by Boyden (1979)[31] is exemplary for the efficiency with which it incorporates social, cultural, economic, technological and especially psychological interactions. From the Figures 1.1 and 1.2 emanates the action by humans at various levels in various conditions. The same is depicted in Table 1.2 below (Bernhard Glaeser (1995).[32]

The model at Figure 1.1 helps in studying and understanding human and environmental impacts caused due to the chain of inter-relationship factors. The non-human factors on the left hand side of the figure are linked to each other, and have their individual and collective links with the human factors on the right hand side, which again are inter-linked through history, culture and social environment. Action at one place has its resultant reaction at one or more factors. Positive human actions add to the health of other factors, while negative actions may start a chain of degeneration.

Figure 1.2 is an integrative model for ecological studies of interactions within human settlements. The main concern of the model is with the impact of societal conditions on the natural environment, and on the quality of human experience. It is a reflection of the explicit and intrinsic values for the integrity of ecosystems and the health and well being of people. The model demonstrates that the state of human health is directly related to humans' interactions with the environment.

Table 1.2 Levels and Types of Action

Level of Action	*Type of Action*
Theory	Knowledge
Aesthetics	Perception
Ethics	Justification of norms
Politics	Implementation of norms
Planning	Regulation
Economy and Technology	Production
Spirituality	Worship and religious fervour

SECTION-IV
DETERMINANTS OF HUMAN ECOLOGY

In recent years, human ecology has been given greater attention by scientists and social scientists. A number of different conceptual approaches have been employed in the study of human ecology. Anderson (1977)[33] argues that the necessity to view man within the framework of his habitat tended towards the adoption of two fruitless positions that for long dominated the thoughts of social scientists. With some simplification these positions can be seen as extremes on a continuum, one pole being environmental determinism and the other cultural determinism. Their less extreme versions are known by the terms "environmentalism" and "possibilism".

Above views, summarise Vayda and Rappaport (1968)[34] tend to separate man and his culture from environment, and behaviour from ecology. They tend to treat them as opposing entities. These views although much diluted, still have their adherents. The major conceptual approaches merit to be discussed independently.

Environmental Determinism or Environmentalism

During the first quarter of the twentieth century many scientists adhered to the doctrine of environmental determinism. These scholars believed that the physical environment, especially the climate and terrain was the active force in shaping cultures, emphasizing that humans were essentially a passive product of the physical surroundings. As analyzed by Jordan and Rowntree (1990)[35], the logic of the 'deterministic view' considered humans equivalent to clay, to be moulded by nature. Similar physical environments were likely to produce similar cultures. Environmental determinists thus viewed human ecology as a 'one-way street'.

Human Ineractional Web

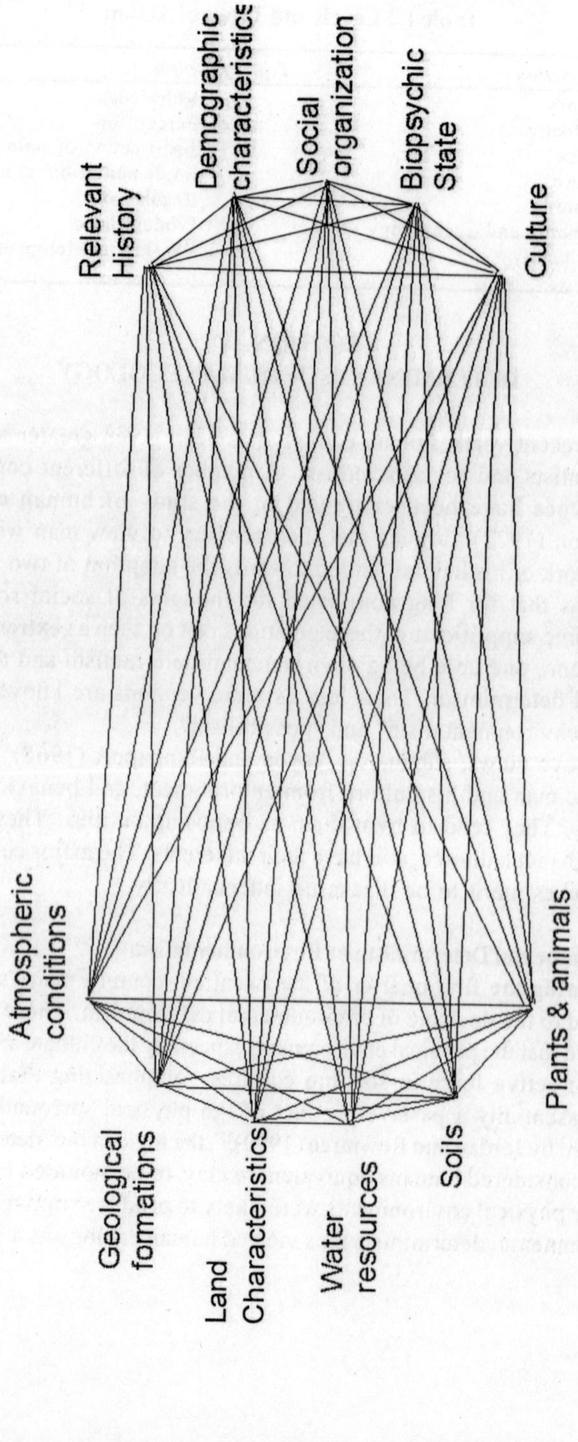

Fig. 1.1 Interactions between the Man-Enviornment system.

Source : Philips-Howard (1985).

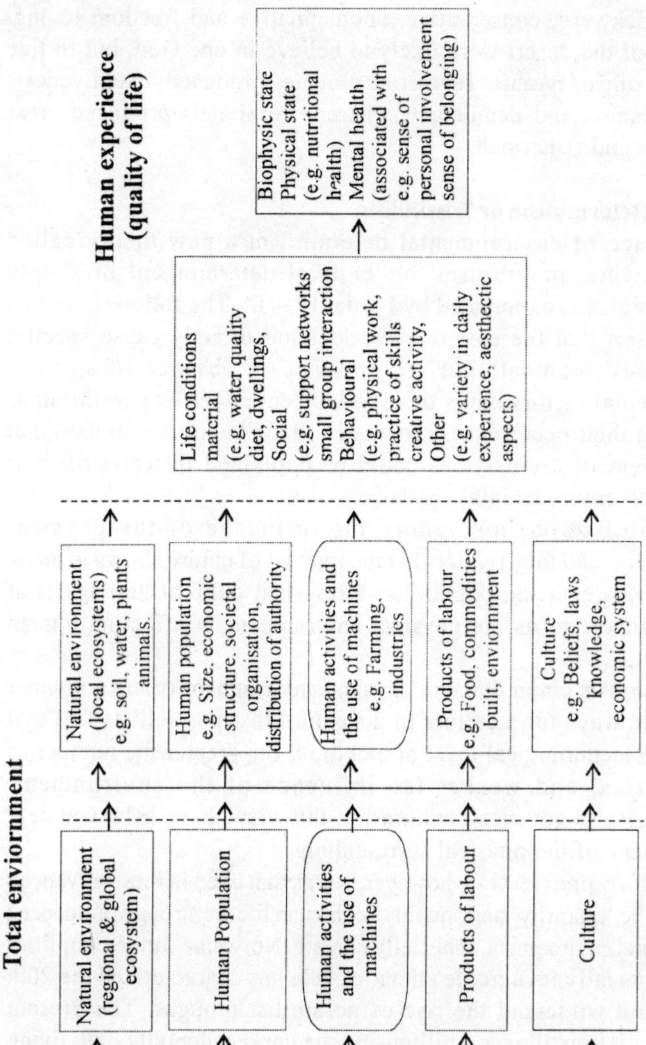

Total enviornment
Natural environment (regional & global ecosystem)
Human Population
Human activities and the use of machines
Products of labour
Culture
Natural environment (local ecosystems) e.g. soil, water, plants animals.
Human population e.g. Size, economic structure, societal organisation, distribution of authority
Human activities and the use of machines e.g. Farming, industries
Products of labour e.g. Food, commodities built, enviornment
Culture e.g. Beliefs, laws knowledge, economic system
Life conditions material (e.g. water quality diet, dwellings,
Social (e.g. support networks small group interaction
Behavioural (e.g. physical work, practice of skills creative activity,
Other (e.g. variety in daily experience, aesthectic aspects)
Human experience (quality of life)
Biophysic state Physical state (e.g. nutritional health)
Mental health (associated with e.g. sense of personal involvement sense of belonging)

Majid Hussain (1994)[36] and Jordan and Rowntree (1990)[37] give a number of examples of determinist beliefs. Determinists believed that people of the mountains were predestined by the rugged terrain to be simple, backward, conservative, unimaginative and freedom loving. Dwellers of the desert were likely to believe in one God, but to live under the rule of tyrants. Temperate climates produced inventiveness, industriousness and democracy, whereas coastlands produced great navigators and fishermen.

Cultural Determinism or Possibilism

In place of environmental determinism a new theory called 'environmental possibilism' or 'cultural determinism' or simply 'possibilism' was expounded by Forde (1934)[38]. The followers of this school assert that the environment does not directly cause specific cultural developments but the presence or absence of specific environmental factors limits the development either by permitting or forbidding their occurrences. Toynbee (1955)[39] is of the opinion that development of civilisations could be explained in terms of their response to environmental challenges.

Possibilists do not ignore the influence of the physical environment, and they realize that the imprint of nature shows in many cultures. However the possibilists stress that cultural heritage is at least as important as the physical environment in affecting human behaviour.

Possibilists claim that any physical environment offers a number of possible ways for a culture to develop. Most possibilists feel that higher the technological level of a culture, the greater the number of possibilities, and weaker the influence of the environment. Technologically advanced cultures, in this view, have achieved near total mastery of the physical surroundings.

Jim Norwaine (1981)[40], however, warns that even in these advanced societies the 'quantity' and 'quality' of human life are strongly influenced by the natural environment, especially climate. Norwaine further amplifies that an unusually favourable climatic cycle has characterized the 20th century and witnessed the rise of possibilist thought. The present world population of over 6 billion and the unprecedentally high living standards typical of advanced countries may be untenable when a deterioration of the weather regime occurs. Perhaps in our zeal to discard environmental determinism, we have overstated the possibilist view.

Economic Deterministic View

This view originated in the West as a branch of the possibilistic view. It is based on the basic ideology of man's mastery over environment, and continued economic and industrial expansion through the application of modern technologies. "The basic thesis of the growth (affluence) school is that economic growth is required for political, social and economic stability." (Savindra Singh,[41] 1991) . 'The quality of environment' in this view normally assumes lower priority in formulating planning proposals and in long-term planning because the deterioration of the environment is generally protracted and socially less oblique than a deterioration in the economy (Park[42] 1980).

Economic determinism is based on two assumptions: -

(a) There is a positive correlation between the population of a given region and the level of economic development and activity in that region.

(b) The interactions of people, resources and society are governed by universal economic principle. (Zelinsky,[43] 1966)"

Based on these two fallacious assumptions, the 'economic determinism' believes in man's ability to solve environmental problems arising out of continued economic growth and industrial expansion. It may be pointed out that this extreme concept of economic determinism of the rich materialistic society led to rapacious exploitation of the natural resources in the Western developed countries, and their colonies in the Third World Countries, and thus has created most of the environmental and ecological problems of global dimension.

Ecosystem based Model

Since culture and environment are dynamic, a need for a more dynamic model was felt. Vayda and Rappaport[44] (1968) proposed an ecosystem based model for human ecology. In their view, human beings constitute simply another population among the many populations of plant and animal species that interact with each other, and also with the nonliving components of their local ecosystem. Thus the ecosystem, rather than the culture constitutes the fundamental unit of analysis in the conceptual framework of human ecology. The cultural traits are of interest only because they can be shown as contributing towards the population's survival in the context of the ecosystem.

The Systems Model

A major scientific development in recent years has been the theory of "general systems theory" which is concerned with the general properties of the structure and functions of systems as such, rather than with their specific components. Odum (1971, 1977)[45] presented a systems view of ecology with particular emphasis on the integrity of the ecosystem as an analytical unit. He also presented mathematical models for analyzing the flow of energy, materials and information between social and ecological systems.

In the systems model of human ecology, both the social systems and the biological systems interacting with each other retain their integrity as ecological systems. In this process each system changes its structural configuration according to its internal dynamics. At the same time each system received energy, material and information from the other and these inputs also influence its structure and functioning. However, each system is also open to influence from other systems of the same kind so that a social system is altered by inputs received from a neighbouring social system just as an ecosystem may be changed by inputs from other ecosystems eg. migration and colonisation.

Nathawat (1995)[46] amplifies on the model of given basic structural and functional relationship in the systems model and has emphasized the following four relational aspects: -

(a) **Inputs from natural ecosystems into the social systems:** The inputs may be in the form of flow of energy (fuel, food, petroleum), materials (building material) or information (sound, visual, stimuli etc).

(b) **Inputs from the social systems into the natural ecosystem:** These may be flow of energy (heat or pollutants), mineral (waste) and information generated by human activities.

(c) **Change in the institution of social system in response to inputs from ecosystem:** These changes may be primary because when an increase in death rate due to environmentally transmitted diseases takes place, it changes the population structure of the society. They may be secondary, when they come from a change generated in other social system. Social system changes in response to inputs from ecosystem are usually adaptive, i.e. they contribute to the continuing survival of the social system under changed environmental conditions.

(d) **Change in the ecosystem in response to inputs from the social system:** These changes may be primary, such as elimination of a species (of biodiversity) by over exploitation or secondary due to alterations in the other ecosystem components caused by primary anthropogenic change in one of its components.

Perhaps the greatest virtue of the systems model of human ecology is that it offers specific guidelines for doing research on human interactions with the environment. The real values of human ecology lies in helping humans to see previously unrecognized relationships between what people do to the environment, upon which they are dependent for their survival.

SECTION - V

AIM AND OBJECTIVES, HYPOTHESIS, METHODOLOGY AND THE CHAPER SCHEME

Aim and Objectives

The main objective of the present work is to study and analyse all facets constituting the human ecological aspects of Upper Rangit Basin, an important mountain enclave in the Eastern Himalayas within the state of Sikkim. The main objective could further be split under following sub-heads and sub-groups:-

1. To assess and evaluate human ecology of the various human and geographic ecosystems in Sikkim.
2. To study the consitutent ethnic groups in Sikkim in general, and Upper Rangit Basin in particular in their traditional human habitat.
3. To assess the determinants of Human Ecology and human ecosystems in the study area.
4. To study and assess the geographical, geological and climatic conditions having overriding effects on human habitat in the Upper Rangit Basin.
5. To analyse the existing terrain features in Upper Rangit Basin with reference to human ecology and habitats that incorporate base line vegetation, land cover, agriculture and fallow lands and vegetation zones.
6. To draw out a human geographical mosaic of Sikkim as a whole with reference to location, history, demography, dress patterns, ornaments and jewellery, food habits, religions, social structure, housing patterns, festivities and political setup.
7. To study the realms of spiritual and deep ecology and the dimensions of *ecodharma* in Sikkim.
8. To study the monasteries with reference to their architecture and the relations of architectural symbolism with human eco-systems and deep ecology.
9. To study the diffusion of Mahayana Buddhism, Hinduism and Animistic faiths of different religious groups in Sikkim.

10. To study the symbolism of prayer wheels, prayer flags, and the wheel of life and what these symbols mean to the local residents, including their sentimental attachments to the symbols in day to day life.
11. To analyse the essence of *chortens* the relevant rituals with regards to passing away of the *Tathagatas*, what the *chortens* mean to the people, and the human and deep ecological relevance to the residents of Sikkim.
12. To bring out the essence of non-violence with respect of Sikkimese Buddhists, Hindus and followers of Animist faiths.
13. To relate the historical facts of Sikkim leading to the Upper Rangit Basin that became the spiritual Centre in the Himalayas in general and Sikkim in particular through the density of religious centres.
14. To dovetail and relate the thought processes of Sikkimese in accordance with their economic and class structures in relation to national culture and their local identity.
15. To study the festivals and other cultural activities in Upper Rangit Basin with special reference to rituals, constructional aspects and invasive aspects of popular culture.
16. To carry out an investigation on spirit possession, shamans and *jhakris* with respect to all ethnic groups viz Bhutias, Limbus and other Nepalese groups and Lepechas.
17. To study aspects of cross-cultural ethnobotany in relation to human ecology in the study area to include medicinal plants, home gardens, floristic diversity, dependent animal diversity and non-timber uses of forest produce.
18. To survey local ethnic weekly village folk markets to assess use of local resources, patterns of specific commodities, thereby evaluating the sustainability of the natural resources, and the factors responsible for each activity.
19. To study the dietary ethnobotanical culture of inhabitants of Upper Rangit Basin including preparation and consumption of traditional fermented and non-fermented foods.
20. To study the human ecology of eco-tourism with respect to areas attracting tourists, the nature trails, spiritual tourism and the constructive and invasive aspects of tourism in Upper Rangit Basin.
21. To draw out a holistic view of all human ecological aspects relating to Upper Rangit Basin and offer recommendations where required as a result of study of interrelationship between all constituents-humans, biotic and abiotic.

Hypothesis

The following hypotheses have been formulated for the present study:

1. The residents living in the secluded habitat of Upper Rangit Basin, which on account of its isolation could easily be termed as an 'Enclave', have continued to live and function in a traditional life style in accordance with their respective ethnic and religious life styles.
2. There has traditionally been a symbiotic relationship between the humans and nature in the harsh and extreme climatic and terrain conditions.
3. Religious faith among the local folk has been the mainstay, and living beacon in their day to day life, festivities and rituals, even at the cost of personal and state development projects.
4. Shamans, *Jhakris* and faith healers have an important place in the society and people follow them in and all ethnic groups.
5. The traditional ecological knowledge has been passed on from generation to generation, and the people have a good knowledge of existing natural floristic life and fauna.
6. Lack of competition due to job reservations has led to sub-par educational standards.
7. Near absence of urban market forces residents to subsist on live stock, home gardens and ethnic weekly folk markets.
8. The virtues of 'nonviolence' through Hinduism and Buddhism and strong traditional ethnic linkages have developed patience and respect for all life form amongst the population.
9. The lamas and monks reign supreme, and the simple ethnic folk consider the lama's word as 'law'.
10. There is a galore of dogmas, superstitions and taboos amongst the simple folk of Upper Rangit Basin.

Chapter Scheme

It has been attempted to adopt the scheming of chapters, in a graduated manner, commencing with introduction to the study area, followed by description of various facets concerning the aspects of human ecology in the Upper Rangit Basin of Sikkim. The study has been split up into eight chapters.

Chapter 1. This is the introductory chapter highlighting the problem, followed by theoretical descriptions on human ecology, human ecosystems, human interactions and the determinants of human ecology. The aims and objectives of the study, methodology and the outline of chapters are also given here.

Chapter 2. Chapter 2 analyses the geological, climatic, landuse and geographical relief profile of the region. Further, the human mosaic of Sikkim with reference to habitat, dressing, food habits, social environment agricultural implements, constituent religious groups and local polity are described.

Chapter 3. This chapter pertains to the realms of spiritual and deep ecology and the dimensions of *eco-dharma* in the study area. The emphasis is on interrelationship between religion followed by inhabitants with natural surroundings. The religious diffusion, monastic architecture, and deep ecology depicted through *chortens* find a detailed description here.

Chapter 4. The aspects of cultural ecology are a direct fallout from the spiritual following and religious faiths. Cultural ecology is the central theme of this chapter. The historical inputs that have lead to current cultural status have been presented. There is a brief description on why the Sikkimese think behave and act, the way they do. There are case studies on celebrations of special festivals, and on shamans and *jhakris* including folktales pertaining to *jhakris* in the Upper Rangit Basin.

Chapter 5. This chapter covers cross-cultural ethnobotany as practiced by the local inhabitants. The major aspects covered are the paradigms of ethnobotany and its human ecological connections, the medicinal and non-timber uses of floristic life, nutritional habits, the home gardens within the area, and the ethnic folk markets. A data presentation on ethnobotanical resources is also given at the end of the chapter.

Chapter 6. Eco-tourism with special reference to nature and monastic spiritual centres is the central theme of this chapter. The role of tour operators, local community and the government have been studied and suggestions offered for planning, monitoring and marketing of tourism industry, while ensuring sustainability of resources at all times.

Chapter 7. A holistic view of human ecology and human interactions through combinations of all facets given in first six chapters has been assimilated here. The behavioural patterns of inhabitants under various situations are realistically defined.

Chapter 8. This chapter deals with recommendations from the point of view of human ecology, and sustainability of human and nature's health through certain suggested actions for inhabitants, visitors, bureaucrats and the government.

Select bibliography on the subject and index are presented at the end.

Plan of Work and Methodology

In order to study the aspects of human ecology of the residents of Upper Rangit Basin, an enclave in the Sikkim Himalayas, various methodological means were adopted. The present work deals with empirical investigations, theoretical study as well as historical inputs. The government policies, both, past and present have had a profound impact on the human ecological aspects, necessitating deep insight into government ministerial reports and departmental statistics. For the purpose of meaningful investigations, structured questionnaires were prepared, and answers were sought through direct interviews and field observations. During the course of the research study, almost the entire period of stay was spent in the study area itself, in order to have a continuous contact with the land and its people.

The methodology followed in the study of the Basin is summarized as under :-

(a) Information concerning the physical environment was facilitated through published materials, maps, charts and constant visits through vehicular and foot travel through the length and breadth of Sikkim.

(b) The information of spiritual and cultural aspects was sought through personal contacts with the concerned people. Visits to monasteries and regular interactions with the lamas and monks helped in understanding and elucidating the aspects of religious ecology. A proper detailed field survey of Tashiding monastery and its surrounding terrain was carried out by staying at a tented camp site at Tashiding itself. Relief details were studied through maps and ground reconnaissance, facilitating pinpoint location of each *gompa*, residential places and *chortens* in Tashiding. Similar study was carried out on the architectural aspects of the monastries and *chortens* of Ralang and Pemayangtse.

(c) Study of cultural ecology was undertaken through the cultural and festival calendar of Sikkim. The visit to Khecheoplari lake was undertaken during the Bumchu festival in 1998. Similar visits were undertaken to the homes of Shamans, *jhakris, yumas, yebas and sambas*.

(d) Ethnobotanical study required a detailed collection and study of research and theoretical material, followed by visits to home gardens at the confluence of Rathong chu and Rangit chu. A field camp and survey in Yuksom forest helped in collection of ethnobotanical data in the surrounds of Yuksom floristic diversity. As for traditional food habits, a visit was undertaken along the Yuksom-Tsoka-Dzongri trails. Preparation of traditional foods was witnessed within the kitchens of residents almost at all places in the study area.

(e) Eco-tourism has arrived in a big way in the Basin, its main centre being the Yuksom-Dzongri area and the Monastic centres. A direct contact was made with the tour organisers, hoteliers, trail porters and the tourists themselves. The government and non-governmental organisations concerned with tourism industry were rich resources of material and statistical details.

The District commissioners of South Sikkim and West Sikkim, the doctors in the PHCs and the District Forest Officers offered help, and facilitated study of official documents. Police department at Ravangla offered details of disciplinary and law and order aspects in the Ravangla sub division in South Sikkim.

Findings from each study and all field observations are given in the respective chapters.

SECTION - VI
REVIEW OF LITERATURE

General

Nearly all the disciplines and professions, in both, the sciences and the humanities are eager to find a common meeting ground in the arena of human ecology. The mosaic of human ecology is built around six themes: Cultural region, Cultural history, Cultural diffusion, Cultural ecology, Cultural integration and Cultural landscape. These themes are applied to a variety of topics concerning human ecology in a given areas demography, habitat, religion, language, ethnicity, politics, folklore, industry and popular culture. A brief review of literature on these approaches is therefore inescapable, and is presented in order. It is an integrative background for ascertaining the holistic emergence of the status of human ecology in the area.

Theory of Human Ecology

One witnesses a surge in literary works dealing with human ecology, social ecology and deep ecology since the late seventies of the twentieth century. John Bennett (1976)[47] frames themes and comparisons on ecological transitions, adaptations, human behaviour and cultural, social and systems aspects of human ecology. Murray Bookchin (1996)[48] portrays the interrelationships between history, civilisation and progress in his edited work "The philosophy of Social Ecology". By far one of the most comprehensive work may be attributed to Bernhard Glaesar (1995)[49] , in his book "Integrated Policy through Human Ecology" where he conceptualises and delimits the subjects, and prepares systems models for application on environment, development and agriculture.

Travelogues

Early works describing the nature of Sikkim society and physical features are in the form of travelogues, and government reports. One of the earliest comprehensive works, encompassing most ingredients of physical features and human ecology was by the eminent botanist Sir Joseph Dalton Hooker[50] . In his book, "Himalayan Journals: Notes of a Naturalist in Bengal, The Sikkim and Nepal Himalayas, The Khasi Mountain, and C,"(1855), Hooker addressed topics such as natural terrain features, flora, history, ethnic groups, cultural and behavioural attitudes, and aspects of spiritual ecology.

Earliest works with special reference to Rangit Basin were addressed to by Major Sherwill (1862)[51] . Dr Mallet (1875)[52] , Colonel G.B Mainwaring (1876)[52] , L.A Waddell(1891)[54] , (1899)[55] , J.C White (1909)[56] and C.G. Bruce (1910)[57] .

Major Sherwill surveyed and explored the base of Kanchendzonga massif, including Rathong glacier and Rathong valley, a major feeder of the Rangit. He described his findings in the "Journal of A Trip Undertaken to Explore the Glaciers of Kanchendzongha Group in the Sikkim Himalaya in November 1861". Dr Mallet dealt with and gave a detailed account of geology and mineral resources in the Rangit Basin area. Colonel Mainwaring and Waddell entered the linguistic fields. While the former wrote a book on the "Grammar of the Rong (Lepchas) Language", the latter wrote an article on the "Place and River Names in the Darjeeling District and Sikkim", which included meanings of names of mountains, passes, rivers, villages and monasteries, alongwith their Lepcha, Bhutia, Nepali and English equivalents. J.C White penetrated into the stupendous mountain valleys of the Rangit and Tista, while serving as the First British Political Officer of Sikkim, and wrote his geographical, historical and personal experiences during various expeditions in his book "Sikkim and Bhutan". Major C.G.Bruce gave detailed descriptive accounts of Dzongri and Pemayangtse Monastery in his work "Twenty Years in the Himalayas".

Published in 1894, H.H Risley edited the first and only" Gazetteer of Sikkim" [58] . The gazetteer gives detailed historical, geographical, botanical and zoological accounts of Sikkim. Spiritual ecology of the *gompas* and monasteries finds a detailed description in the gazetteer.

Taking a holistic view of human ecology, not withstanding their year of publication, the subsequent literary works can be best reviewed by separate analyses of works on various themes.

Works on Ethnic Groups

Sikkim drew attention of anthropologists and ethnologists from Western countries and the Orient. In the 1930's John Morris [59] and Geoffery Gorer [60] spent several months in Dzongu area of North Sikkim to study the Lepchas. Their research works were published in books, "Living with Lepchas: A book about the Sikkim Himalaya" by John Morris (1938) and "Three months in a Himalayan Village : An Account of the Lepchas of Sikkim" by Geoffery Gorer (1976). Both authors deal with the social, political, cultural and spiritual ecological aspects.

R.N Thakur in his book "Himalayan Lepchas" [61] (1988) explored the customs, manners, food habits and other social institutions amongst the Lepchas, with an emphasis on the process of change in their life style due to invasive aspects of popular cultural variety. An inside view of the Lepchas is presented by A.R Foning[62] (1987) in the book "Lepcha : My Vanishing Tribe". Other authors who have addressed research works to Lepchas include Siiger Halfden[62] (1972) and K.P Tamsang[65] (1983).

K.S Singh[66] (1993) edited 39th volume of Anthropological Survey of India, "People of India : Sikkim", which offers a profile of almost all 25 communities identified in the state. Tanka B. Subba[66] (1989), in his work "Dynamics of a Hill society: The Nepalis in Darjeeling and Sikkim Himalayas", has a deep look into the history of individual Nepali castes, and describes their cultural traits. A large detailed volume, "The Limbus of the Eastern Himalayas: with special reference to Sikkim" has been authored by J.R Subba[67] (1999).

A prominent work on the Bhutias was published by Saradindu Basu[68] (1966), who carried out his research in North Sikkim, where the Bhutias are the majority ethnic group. A social and political study was carried out in his Ph.D research by a sociologist A.C Sinha[69], who prepared his research document on the "Elites of Sikkim: A study in Political Development", (1972). T.B Subba[70] (1999) worked on political and cultural ethos with special reference to current status of Kirata tribes in Sikkim and Nepal in his work "Politics of culture. A study of three Kirata communities in Eastern Himalayas".

Folklores and Shamans

Folklores are an important ingredient in the study of human ecology. C.De. Beauvoir Stocks[71] undertook a research on the customs and folklores on particular lines of Lepcha life during 1925. George Kotturan[72] collected and published, "Folk Tales of Sikkim" in 1989, some of which highlight the Sikkimese love for nature and indigenous fauna and flora.

Shamanism, spirit possession and mystical rites are performed and openly practiced in Sikkim and Nepal Himalayas. John T. Hitchcock and Rex L. Jones[73] have jointly edited a work on the spirit possessions, which most importantly describes the reincarnation possession, and other classifications of shamans and *jhakris*.

Tulsiram Sharma 'Kashyap'[74] (1988) in "Sikkim : A Himalayan Realm" has depicted Sikkim's history, folklore culture, its nature and also political developments in the form of poems, through the intimate interaction between two characters Lakpa and Rajani, both belonging to different ethnic groups.

Geography and Habitat

The most detailed and authentic work on the human habitat and geography of the Rangit Basin however, can be credited to Maitreyi Barua[75] (1988) in her Ph D thesis Titled, "Habitat and Economy in an Eastern Himalayan Enclave: Case study of the Rangit Basin'. She has given a detailed account of the inhabitants, human geography, and agro- economy alongwith details of all physical terrain features, supported by suitably scaled maps of the Rangit Basin. Again in 1997,Maitreyi B. Choudhary[76] published a paper through Centre for Himalayan Studies (CHS), North Bengal University, on Rathong Chu hydro projects imbroglio in the Upper Rangit Basin. Maitreyi Choudhary's article in CHS journal Himalayan Miscellany[77] (1997) deals with the environment and human adaptations in the Rangit Basin.

Spirituality and Culture

Spirituality and religious realms have major influence on the thinking, actions and functional ethos of a community, that affect the human eco-systems and human ecology of the residents. Lama Kazi Dawa-Samdup[78] (1935) in his work (Edited by Evans Wentz) "Tibetan

Yoga and Secret Doctrines" explains the principles, methods and wisdom contained in all categories of Mahayana Yogic precepts. Hajime Nakumara[79] (1964) describes why the Indians and Tibetans (Sikkemese being a part of the diffused culture) do and behave the way they do, in " Ways of Thinking of Eastern Peoples". The ecological aspects of Buddhism and Hinduism are well illustrated through anecdotes in "Hinduism and Ecology", and "Buddhism and Ecology" by Ranchor Prime[80] (1992) and Martine Batchelor and Kerry Brown[81] (ed) (1992) respectively. David Brazier[82] (1997) describes the Buddhist psychology of character, adversity and passion in book "The Feeling Buddha".

CHS staff have from time to time carried out research on the theme of religion, and monastic studies on Sikkim. Ranju R. Dhamala[83] (1993) and D.P. Boot[84] (1996) produced detailed papers on the geographical location and economic structure of monasteries in Sikkim. Karubaki Dutta[85] (1997) published a research paper on "Inter Ethnic Relations in Sikkim in Historical Prospective", and another paper dealing with religion, on "Changing Status of Buddhism in Sikkim" [86] (1997). Karubaki Dutta and Tanka B. Subba published and edited the book, " Religion and Society in the Himalayas" [87] (1991), containing articles on monastic studies, bibliographical study, and functional *karmic* ethos in Lamaist society.

Sikkim Research Institute of Tibetology, Gangtok (SRIT), published presentations made in national seminar on "Guru Padmasambhava's Contributions to Sikkim" [88] (1995). The spiritual ecology theme of the *Chortens* (stupas) has been discussed in details in SRIT bulletin of Tibetology[89] (1970) and (1969) [90] through articles by Lama Anagarika Govinda and B. Ghosh respectively. An all inclusive work on "Stupa and its Technology" was carried by Pema Dorjee[91] (1996).

Rumtek in West Sikkim is the famous monastery of the Kargyud Sect, and is also the seat of the Karmapa. Ralang located in the Upper Rangit Basin has the oldest monastery of the Kargyud Karmapa sect in Sikkim. The book "Karmapa Khenno - A Sikkemese Point of View" [92] (1993) gives a detailed account of the tradition of reincarnation of Karma Kargyud Lineage. It may be noted that the sudden arrival of the 17th Karmapa to Dharamsala in January 2000, became the main news items in national and international media.

Ethnobotanical Studies

Interrelationship between earliest inhabitants of Sikkim with the fauna and flora of the place has continued since the era of gathering and hunting. Plants have also been extensively used for medical purposes. Upper Rangit Basin is known for its rhododendron trails. Udai C. Pradhan and Sonam T. Lachungpa [93] have traced all rhododendron trails, and explained aspects of its ecology, conservation and uses in their book, "Sikkim- Himalayan Rhododendron" (1991). Topdhen Rai and Lalit Kumar Rai [94] (1994) have given a compendium of Sikkim trees renowned for their resplendent floral wealth in boom, "Trees of the Sikkim Himalaya". The medicinal plants of the Sikkim Himalaya, their status, usage and potential find a detailed description by Lalit Kumar Rai and Eklabya Sharma[95] (1994) of G.B. Pant Institute of Himalayan Environment and Development. Arvind Saklani[96] (1994) gathered information and data in the book "Cross-cultural Ethnobotany of Northeast India". It contains details and usage of flora found in Sikkim. Consumption pattern of traditional fermented and non fermented foods have been elaborated upon in details by J.P. Tamang[97] (1998) and H. Youzan and J.P. Tamang[98] (1998).

Ecology and Conservation

A study was directed at ecology, and changing contours of Sikkim culture by Veena Bhasin[99] (1989). Later, working with Man and Biosphere Programme (MAB) of the UNESCO, M.K. Bhasin and Veena Bhasin jointly authored "Sikkim Himalaya : Ecology and Resources Development" [100] (1995). Jack D. Ives and Bruno Messerli of the United Nations University analysed the environment of the Nepal and Sikkim Himalayas in book, "The Himalayan Dilemma : Reconciling Development and Conservation" [101] (1995).

Polity, Education and Economy

Literary research relating to politics, education economy and development has been popular in the past decade. Books by Manas Das Gupta[102] (1992), Jigme N. Kazi[103] (1993), N. Sengupta[104] (1985) and by Suresh Chand Rai, Rakesh Sundriyal and Eklabya Sharma[105] (1998) are of note. A study team of J.N. University under Mahendra P. Lama is currently preparing the "Sikkim Human Development Report (1999), (Unpublished)[106].

Summary

This review of literature reveals that plenty of literary works are available on Sikkim. Literature on human ecology with a holistic panoramic view is however scarce. A void is also seen by the absence of literary works on the human ecosystems of the Upper Rangit Basin.

It being the original seat of the first Chogyals, and it being the abode of the original and historical spiritual centres and monasteries of Sikkim, a deeper research into its human ecology is called for.

References

1. Bernhard Glaeser, *Environment Development Agriculture : Integrated Policy through Human Ecology*, UCL Press Limited, University College, London 1995.e
2. Sigmund Kvaloy, in Inside Nature in *The Future of Progress* by Helena Noberg- Hodge (Edited), Natraj Publishers, Dehradun 1994 (P-125).
3. Terry. G. Jordan and Lester Rown tree, *The Human Mosaic,* Harper Collins Publishers, 1990. (P 276-277).
4. Bernhard Glaeser, *ibid.*
5. Rajiv. K. Sinha (Ed), *Refresher Course on Environment, IGC for HEEPS,* University of Rajasthan,1995.
6. Murray Bookchin, *A Philosophy of Social Ecology*, Rawat Publications, Jaipur, 1996.
7. Ramakant, in Politics of Environment, *Refresher Course on Environment,* IGC for HEEPS, University of Rajasthan, 1995.
8. *Ibid.*
9. Bernhard Glaeser, *ibid.*
10. *Ibid.*
11. Southwick Charles H, *Ecology and Quality of Environment*, Litton Educational Publishing inc, New York 1972.
12. Toffler Alvin; *The Third Wave*, William Collins Sons and Co Ltd, London 1980.
13. Odum EP: *Ecology The Link Between the Natural and the Social Sciences*, Holt, Kinehart and Winston inc, New York 1975 (P 15,16).
14. Terry. G. Jordan and Lester Rown tree, '*The Human Mosaic*', Harper Collins Publishers, 1990 (P 261, 262).
15. Daniel, SC; *Ethnicity, Language and Nationalism : Some Conceptional Issues*. Ethnicity and culture in North East India (edited) MM Agrawal, Indus Publishing Company, New Delhi, 1996.
16. Khan, TI; *Biodiversity Conservation and Sustainable Development*, Pointer Publishers, Jaipur, 1998 (P-126).
17. Andrew Gouldie; *The Human Impact - on the Natural Environment*, Blackwell Publishers, Oxford, UK, 1990.
18. Odum, Eugene. P : Ecology, Holt, Kinehart and Winston Inc, New York, 1975 (P 151).
19. Hawley, Amos. H: *Ecology and Human Ecology* (1944) in 'Selected Articles in Social Ecology' (ed) by Wittman, Jr, James S; Mss Information Corporations, 655 Madison Avenue, New York, Ny, 1973 (P-12).
20. IOHE (ed.), *Human Ecology (brochure)*, International Organisation for Human Ecology, 1981.
21. Klaus, G & M. Buhr (eds) 1976, *Philosophisches Worterbunch* 12th edition, Leipzing; Encyclopedia, 1976.

22. Bernhard Glaeser, *Environment, Development, Agriculture-Integrated Policy Through Human Ecology*, UCL Press Ltd, London, 1955.
23. Anderson, James. N.; *Ecological Anthropology and Anthropological Ecology* in Hand book of social and cultural Anthropology, Johan. J. Honigmann (ed), Houghton Mifflin Company, Boston, USA, 1977.
24. Steward, J.H; Theory of Culture Change, University of Illionois Press, Urbane, 1955.
25. Steward, J.H.; *Cultural Ecology*, In International encyclopedia of the Social Sciences, University of Chicago Press, 1968. (4:337-44).
26. Bates, M ; *The Forest and the Sea : A look at the Economy of Nature and the Ecology of Man*; Random House, New York, 1960.
27. Vayda A.P; and R.A. Rappaport; *Ecology : Cultural and Non-Cultural* in Introduction of Cultural Authropoloty (ed), Houghton Mifflin Press, Boston, 1968.
28. Andy, J.R.; *The Environment in Human Ecology*, Scientific Publication No 123, Washington, D.C. : Pan American Health Organisation, 1965.
29. ____________, *The Ipsefact : In Ecology, Ethology, Parasitology, Sociology and Anthropology* in Behaviour and Environment (ed) A.H. Esser, Plenum Press, New York, 1971.
30. Philips-Howard, K.D ; *A Human Ecological Model for Studying the Impact of Development Project upon Niger Delta* in Mangrove Ecosystem, University of Harcourt, Nigeria, 1985.
31. Boyden, S. ; *An Integrative Ecological Approach to the study of Human Settlements* (MAB Technical Notes - 12), Paris, UNESCO, 1979.
32. Bernhard Glaeser; *ibid.*
33. Anderson, James N. ; *Ecological Anthropology and Anthropological Ecology* in Hand book of Social and Cultural Anthropology, John. J. Honigmann (ed), Houghton Mifflin Company, Boston, USA, 1997.
34 Vayda, A.P. and R.A. Rappaport; Ecology : *Cultural and Non-Cultural* in introduction to Cultural Anthropology (ed), Houghton Mifflin Company, Boston, USA, 1968.
35. Jordan, Terry. G. and Lester Rowntree: *The Human Mosaic*, Harper Collins. Publishers, New York, USA, 1990.
36. Husain, Majid: *Human Geography*, Rawat Publication, Jaipur, 1994.
37. Jordan, Terry. G. and Leaster Rowntree : *ibid.*
38. Forde, C.D.; *Habitat, Economy and Society; A Geographical Introduction to Ethnology*' Dutton, New York, 1934.
39. Toynbee, Arnold J. ; *A study of History* (Volume II), Oxford University Press, London 1956.
40. Norwaine, Jim. ; *Geography as Human Ecology*, in International Journal of Environmental Studies, 1981.
41. Singh, Savindra,; *Environmental Geography*, Prayag Pustak Bhawan, Allahabad, 1991.
42. Park, C.C.; *Ecology and Environmental Management*, Butterworths, London, 1980.

43. Zelinsky, W.: *A Prologue to Population Geography*, Printice Hall, Englewood Cliffs, New Jersey.
44. Vayda, A.P. and R.A. Rappaport; *ibid.*
45. Odum, E.P.; *Fundamentals of Ecology*, W.B. Saunders Company USA, (1971,1977).
46. Nathawat, G.S.; *Refresher Course on Environment*, Academic Staff College, University of Rajasthan, Jaipur, 1995.
47. Bennet, John, W., *The Ecological Transition : Cultural Anthropology and Human Adaptation*, Peragamon Press Incl., 1976.
48. Bookchin, M.,*The Philosophy of Social Ecology*, Rawąt Publications, Jaipur, 1996.
49. Glaeser, B., *Environment, Development, Agriculture: Integrated Policy through Human Ecology*, UCL Press, London, 1995.
50. Hooker, J.D., *Himalayan Journals : Notes of a Naturalist in Bengal, The Sikkim and Nepal Himalayas, The Khasi Mountains & C.2 Vols. Murray*, London, 1985.
51. Sherwill, J.L., "*Journal of a Trip Undertaken to Explore the Glaciers of Kancheningha group in Sikkim Himalaya in November 1861*", Journal of the Asiatic Society of Bengal. XXXI, 1862.
52. Mallet, F.R.,"On the Geology and Mineral Recourses of the Darjeeling District and the Western Duars", *Memoirs of the Geological Survey of India*, Vol XI, Pt I, Calcutta, 1875.
53. Mainwaring, G.B, '*A Grammar of the (Rong) Lepcha Languages*, Calcutta, 1876.
54. Waddell, L.A, "Place and the River Names in the Darjeeling District and Sikkim", *Journal of the Asiatic Society of Bengal*, LX, Pt-I, 1891.
55. __________, *Among the Himalayas*, Constable, London, 1899.
56. White, J.C, *Sikkim and Bhutan : 21 years on the North East Frontier*, 1887-1908 Longmans, New Tork, 1909
57. Bruce, C.G., *Twenty Years in the Himalaya*, Arnold, London, 1910
58. Risley. H.H, *Gazetteer of Sikkim*: (ed) Bengal Government Secretariat, Calcutta, 1894.
59. Morris, , *Living with Lepchas : A Book About the Sikkim Himalayas*, Heinemann, London, 1939
60. Gorer, G, *The Lepchas of Sikkim*, Basic Books, 1967, Reprint Culture Publishing House, Delhi, 1984.
61. Thakur, R.N., *Himalayan Lepchas*" Archives publishers, New Delhi, 1988.
62. Foning, A.R., *Lepchas : My Vanishing Tribe*", Sterling Publishers Private Ltd New Delhi, 1987
63. Siiger, Halfden.,"*Lepchas : Culture and Religions of Himalayan People*", National Museum of Denmark, Copenhagen, 1967
64. Tamsang, K.P., "*The Unknown and Untold Reality about the Lepchas*' Lyangson Tamsang, Hongkong, 1983.

65. Singh, K.S., (ed) 'People of India : Sikkim (Vol XXXIV), Seagull Books, Calcutta, 1993.
66. Subba, T.B., '*Dynamics of a Hill Society : The Nepalis in Darjeeling and Sikkim Himalayas*' Mittal publications, Delhi, 1989.
67. Subba, J.R., '*The Limbus of the Eastern Himalayas - With Special Reference to Sikkim*' SYMS, Gangtok, 1999.
68. Basu, Saradindu, '*Bhots of North Sikkim : Man in India* Vol.46, No 2, Ranchi, 1992.
69. Sinha, A.C., '*Elites of Sikkim : A Study in Political Development*', Ph.D. Thesis, 1992.
70. Subba, T.B., '*Politics of Culture : A Study of Three Kirata Communities in Eastern Himalayas*' Orient Longman Ltd, Chennai, 1999.
71. Stocks, C.De.B., *Sikkim : Customs and Folklore*, Cosmos Publications, reprint, 1975.
72. Kotturan, G., '*Folk Tales of Sikkim*' Sterling publishers Pvt Ltd, New Delhi, 1989.
73. Hitchcock, J.T., and Jones, R.L., (ed)'*Spirit Possession in Nepal Himalayas*' Vikas Publishing House, Pvt Ltd, New Delhi, 1994
74. Kashyap, T.S., '*Sikkim : A Himalayan Realm*' Atma Ram and Sons, Delhi, 1988.
75. Barua, M., '*Habitat and Economy in Eastern Himalayas : Case Study of the Rangit Basin*', Ph. D. Thesis, 1988.
76. Choudhury, M.B., '*Development Project Vis-à-vis Ethno-Religious Sentiments: The Rathongchu Imbroglio in Sikkim*', Occasional Paper-12, CHS, North Bengal University, Siliguri, 1997.
77. _________., 'Environment and Human Adaptation in an Eastern Himalayan Enclave : Case study of the Rangit Basin in the Sikkim Darjeeling Himalaya', The Himalayan Miscellany, Vol II, CHS, North Bengal University, Siliguri, 1997.
78. Samdup, Lama, K.D., '*Tibetan Yoga and Secret Doctrines*', W.Y. Evans - Wentz (ed), Oxford University Press, London, 1935.
79. Nakurmara, H., '*Ways of Thinking of Eastern Peoples*', Motilal Banarsidass Publishers Pvt Ltd, New Delhi, 1964.
80. Prime, R. (ed)., '*Hinduism and Ecology*', Motilal Banarsidass Publishers Pvt Ltd New Delhi, 1992.
81. Batchelor, M. and Brown, K., (ed), '*Buddhism and Ecology*', Motilal Banarsidass Publishers Pvt Ltd, 1992.
82. Brazier, D., '*The Feeling Buddha*', Harper Collins Publishers, India, 1977.
83. Dhamala, R.R., '*Monasteries of Sikkim*', Occasional Paper No.8, CHS, University of North Bengal, Siliguri, 1993.
84. Boot, D.P., '*Monasteries in Sikkim : A Geographical Study*', Occasional Paper No. 9, CHS, University of North Bengal, Siliguri, 1996.
85. Dutta, K., '*Inter-Ethnic Relations in Sikkim in Historical Perspective*', Occasional Paper No. 11, CHS, North Bengal University, Siliguri, 1977.

86. _______, '*Changing Status of Buddhism in Sikkim*' in The Himalayan Miscelleny, Vol II, CHS, North Bengal University, Siliguri, 1997.
87. Subba, T.B., and Dutta, K., (eds), '*Religion and Society in the Himalayas*' Gian Publishing House, New Delhi, 1991.
88. Seminar Paper. *Guru Padmasambhava's Contributions to Sikkim*', SRIT, Gangtok, 1995.
89. Govinda, Lama, Angarika, "*The Historical and Symbolical Origns of Chorten*; in Bulletin of Tibetology, SRIT, Gangtok, Vol VII, No 3, 1970.
90. Ghosh, B., '*Upnishadic Terms in Buddhism*', Bulletin of Tibetology, Vol VI, No. 3, SRIT, 1969.
91. Dorjee, P., '*Stupa and its Technology*', IGNCA, Motilal Banarsidass Publishers, Pvt Ltd, 1996.
92. Zangmo, A.K.D., '*Karmappa Khenno - A Sikkimese Point of View*', Kesangi House, MG Marg, Gangtok, 1993.
93. Pradhan, U.C., Lachungpa, S.T., *Sikkim - Himalayan Rhododendrons*', Primulaceae Books, Kalimpong, West Bengal, 1990.
94. Rai, T., and Rai, L., "*Trees of the Sikkim Himalaya*', Indus publishing Company, New Delhi, 1994.
95. Rai, L., and Sharma, E., '*Medicinal Plants of the Sikkim Himalaya : Status*, Usage and Potential', Bishen Singh, Mahendra Pal Singh, Dehradun, 1994.
96. Saklani, A., & Jain, S.K., '*Cross Cultural Ethnobotany of North East India*', Deep Publications, New Delhi, 1994.
97. Tamang, J.P., '*Indigenous Fermented Foods of the Sikkim Himalaya*' in, Sikkim : Perspectives or Planning and Development, Sikkim Science Society, Tadong, Sikkim, 1998.
98. Yonzan, H., and Tamang, J.P., '*Consumption Pattern of traditional Foods in the Sikkim Himalaya*' in Journal of Hill Research, Sikkim Science Society, II (I), 1998.
99. Bhasin, V., '*Ecology Culture and Change*;, Mittal, M.C, Inter-India Publication, New Delhi, 1989.
100. Bhasin, M.K., and Bhasin, V., '*Sikkim Himalayas : Ecology and Resource Development*', Kamla Raj Enterprises, Delhi, 1995.
101. Ives, J.D., and Messerli, B., '*The Himalayan Dilemma*' U.N. University, Routledge, London, 1989.
102. Gupta, M.D., '*Sikkim : Problems and Prospects of Development*', Indus Publishing Company, New Delhi, 1982.
103. Kazi, J.N., '*Inside Sikkim : Against the Tide*' Hill Media Publications, Gangtok, 1993.
104. Sengupta, N., '*State Government and Politics: Sikkim:*, Sterling Publishers Pvt Ltd New Delhi, 1985.
105. Rai, S.C. et al., '*Sikkim: Perspective for Planning and Development*', Sikkim Science Society, Tadong, 1998.
106. Lama, M.P., '*Sikkim Human Development Report*' Government of Sikkim Awaiting Publication), 1999.

2

The Study Area, Terrain Profile and Geographical Mosaic of Sikkim

SECTION-I
THE STUDY AREA

Introduction

No region in the world excites the imagination and calls up visions of the exotic more than the Himalayas. Their soaring peaks and fertile valleys have nourished some of the world's most ancient cultures and religions. (Maurice Strong, 1988)[1]. The Himalayas have, since ages regulated the climates, and nourished the biodiversity, including the natural and human ecosystems, not only within its peaks, ridges and valleys, but have encompassed almost the entire Indian Subcontinent.

Sikkim is a tiny hill state ensconced in the Eastern Himalayas, sandwiched between Tibet, Nepal and Bhutan. The narrow Darjeeling-Siliguri corridor separates it from Bangladesh. It is a meeting point of multiple cultural groups. Lying between latitude 27 degree N and 28 degree N, and longitudes 88 degree E and 89 degree E, it occupies an area of 7096 sq kms. Almost entire Sikkim state forms the catchment of river Tista, which flows North-South, bisecting the state in two halves. River Rangit a major tributary, originates from the Kanchandzonga massif and flows through West and South Sikkim, till it unites with Tista at the Melli confluence point (Map No 1)

From time immemorial, the Rangit had a strong impression on the life and activities of the people inhabiting its catchment area. Its name is often referred to in somewhat special attention for its individual geographic identity, hydrological potentiality, economic viability and aesthetic beauty. (Maitreyi Barua, 1988).[2]

The Rangit basin is located in the Sikkim- Darjeeling Himalayas between latitudes 27 degree 02 minutes N to 27 degree 33 minutes N and longitudes 88 degree 01 minutes E to 88 degree 27 minutes E. The Mount Narsing-Mainam-Tendong Range delimits the Eastern catchment, while the Singalila Range demarcates the Western boundary (Map No 2). The township of Legship in West Sikkim District virtually divides the Rangit in two halves, the catchment area North of Legship being the Upper Rangit Basin, and to the South being the Lower Basin.

Aim

The aim of this study is to address the human ecology aspects of the Upper Rangit Basin. The study of human ecology being holistic, it was pertinent that the research includes the physical environment as well as the human ecosystems. It was therefore necessary that the physical terrain and climatic characteristics and the human geography are given due weightege in the description of the study area.

Location, Physical and Climatic Features

The Rangit Catchment (zone under study) is located in the South and West districts of Sikkim. The area lies between 27 degree 16 minutes 3 second to 27 degree 32.2 minutes North latitudes and 88 degree 2.5 minutes to 88 degree 25 minutes East longitudes (Map No 3). The catchment area is drained by its tributaries Reli Chu, Rathang Chu, Rangit Chu etc. The rivers generally flow in North-South direction. The terrain is mainly mountainous in nature with steep to very steep slopes. The predominant aspects of terrain are East and West. The elevation ranges from 600m to 7338m.

The climate is cold and humid. Temperature varies between 1 degree C to 32 degree C. There are three distinct seasons viz. summer, rainy and winter. The rains commence during May and last upto late October. Peak rainfall occurs during July and August. Average annual rainfall is 2620.9mm. The average annual rainfall pattern is given at Fig 2.1. The winter season ranges from November to February.

Geology

Pre-Cambrian rocks represented by Khangenedzonga gneiss, Chungthang formation, Darjeeling gneiss, and Rangit schist formation cover a major portion of the Sikkim Himalayan deposition which took place initially in a compound geo-synclinal basin with a sub-merged miogeoanticlinal ridge. The subsequent pre-Cambrian oregeny led to rising of this mio-geoantlelinal ridge dividing the basin into the Northern Eu-geogyncline and Southern miogeosyncline in which deposition of palaeozoic rocks took place. Detached outcrops of post palaeozoic rocks are also present in extreme North. The area suffered three major phases of tectonic movements leading to folding and faulting which gave rise to nappe and sub-nappe structures and formation of release faults.

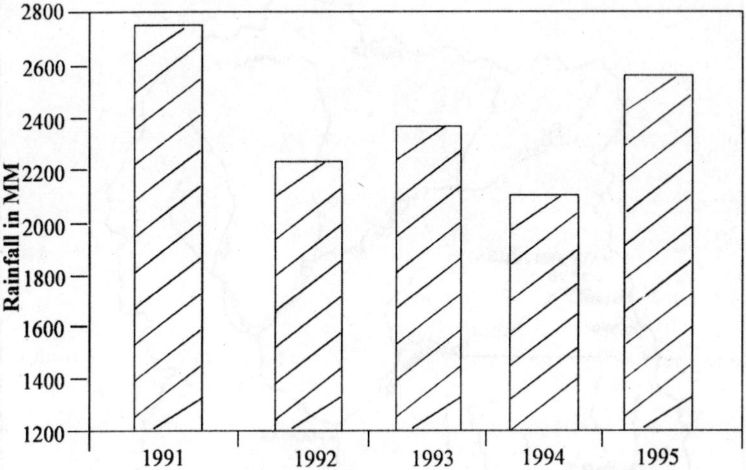

Fig. 2.1
Annual Rain Fall in the Rangit Basin

In the basin pre-Cambrian rocks belonging to the Khangenzonga gneiss group. Chungthang formation group, Darjeeling gneiss group and early Palaeozoic disposed off in the area due to geological structures. Stratigraphic disposition is not actually seen in the field. The tectonic trend in the area is dominantly North-East and South-West but in West it is North-Northwest and South-Southeast. Due to presence of cross faults specially in South-West corner, both the trends are present on either sides.

Daling phyllites are prone to extensive landslides with a slump type failure which is restricted within slopewash material. Landslides do occur in granite gneiss and quartzite but to a restricted scale and are mostly debris avalanches or slide or a rock fall.

Soils

The soils of the catchment are mostly of recent origin and are not well developed. Generally the soils are loamy sand to silty clay loam in

MAP OF SIKKIM

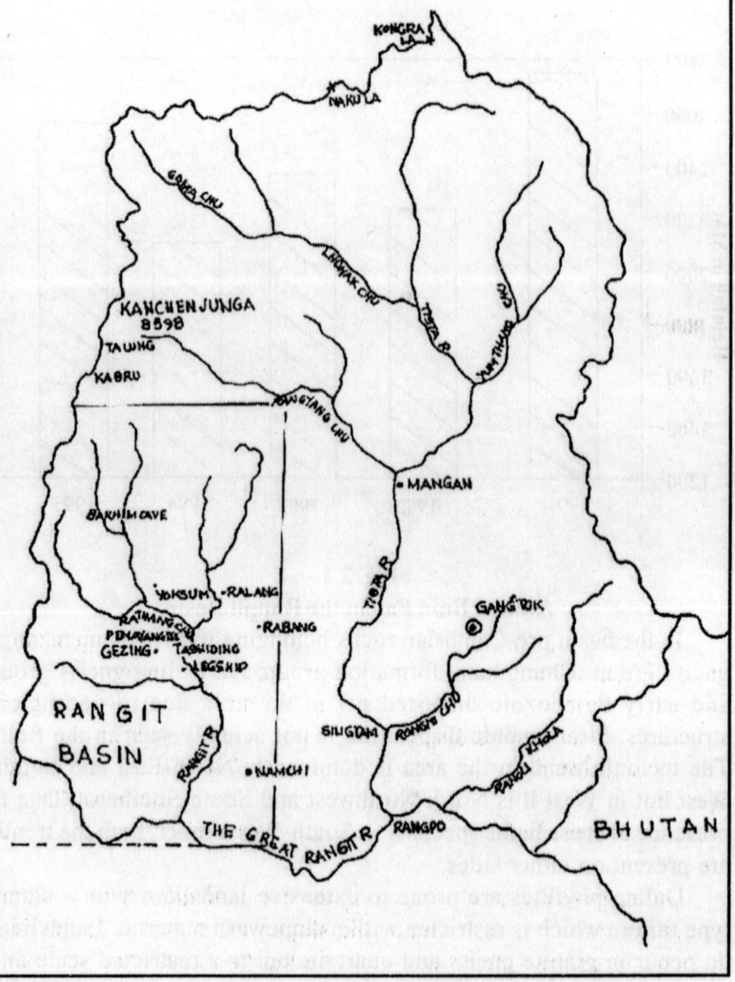

Map - II

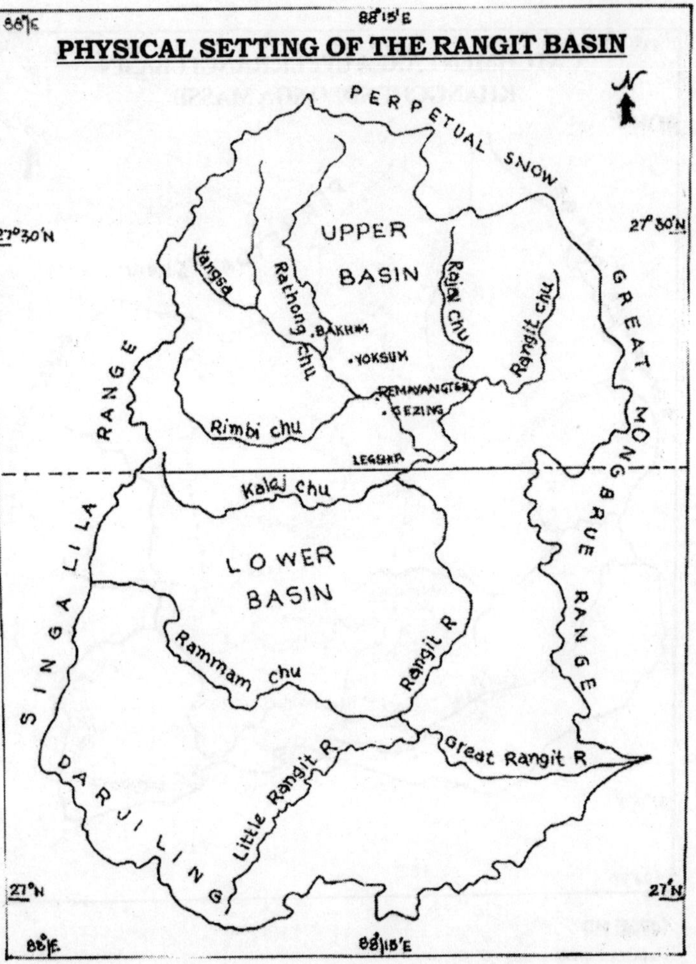
88°E
88°15'E
PHYSICAL SETTING OF THE RANGIT BASIN
N
PERPETUAL SNOW
27°30'N
27°30'N
UPPER BASIN
Yangsa
Rathong chu
Rajai chu
Rangit chu
BAKHIM
YOKSUK
REMAYANGTSE
GEZING
LEGSHIP
Rimbi chu
Kalej chu
GREAT MONG BRUE RANGE
SINGALILA RANGE
LOWER BASIN
Rammam chu
Rangit R
Great Rangit R
Little Rangit R
DARJILING
27°N
27°N
88°E
88°15'E

MAP - III

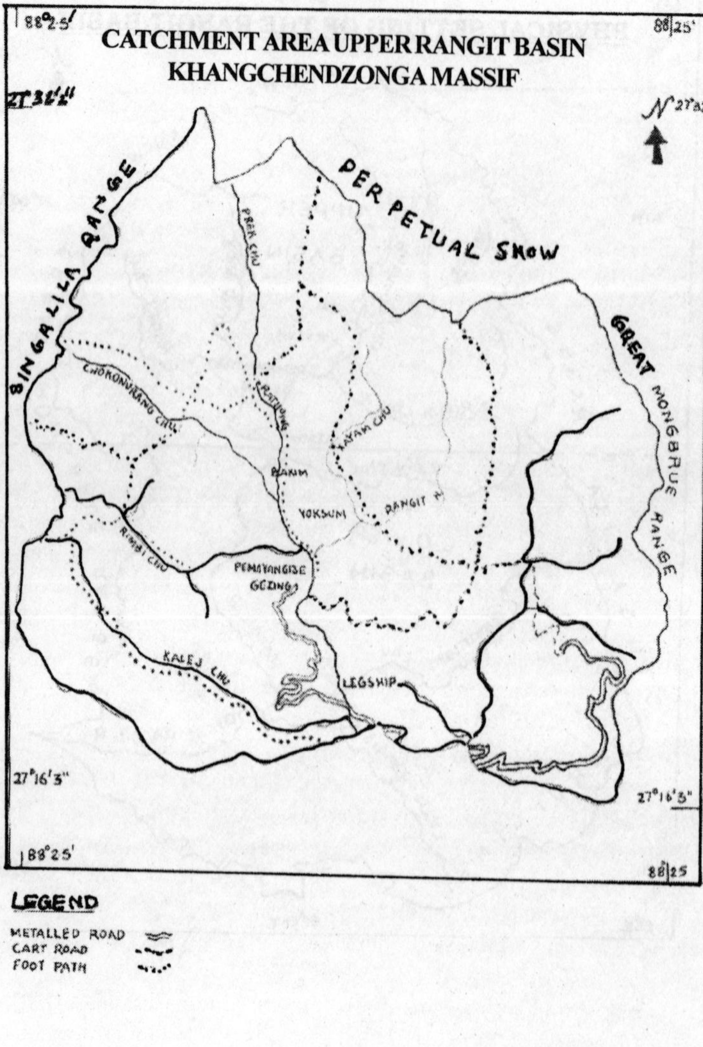
CATCHMENT AREA UPPER RANGIT BASIN
KHANGCHENDZONGA MASSIF
SINGALILA RANGE
PERPETUAL SNOW
GREAT
MONGBRUE RANGE
N
CHOKONVRANG CHU
PREK CHU
RATHONG
TAYAM CHU
RANGIT N
YOKSUM
RUMBI CHU
PEMAYANGTSE
GEZING
KALEJ CHU
LEGSHIP
88°25'
27°16'3"
LEGEND
METALLED ROAD
CART ROAD
FOOT PATH

texture , dark grayesh brown to dark brown in colour, excessively drained to moderately well drained and developed on micaceaus gneissic parent rock and are susceptible to severe erosion hazards due to differences in altitude, slope and rainfall.

Vegetation

The predominant vegetation in the catchment is forests. The agriculture is practiced in low to medium altitudes areas. Scattered scrub and barren rocky areas are also more common. The forests are classified into following major types in accordance with the elevation and all these exist in Upper Rangit Catchment (NHPC Report, 1998)[3] :-

- East Himalayan sub-tropical wet hill forests.
- East Himalayan wet temperate forests.
- East Himalayan mixed coniferous forests.
- Alpine scrub and pastures.

SECTION-II
TERRAIN ANALYSIS OF RANGIT CATCHMENT

General

Upper Rangit river catchment has 26 watersheds covering an area of 979.02 kms. The catchment has basic problems of high run-off and high sediment loads. In addition areas having steeper slopes are being cultivated on partly terraced lands resulting into erosion problems. The biotic interferences such as heavy grazing, construction and widening of roads, mining operations also aggravate erosion levels in the area. The frequent occurrence of land slips, landslides and stream bank erosion also indicate the high run-off and sediment loads in the area. In order to preserve catchment conditions and regulate the run-off/ erosion level in the catchment, the catchment area treatment plans were started in 1995. These watersheds for treatment were selected based on erosion intensity physiography and landcover (NHPC report) and the total area under these six watersheds under treatment is 12,992 ha. The present study was conducted to assess baseline watershed characteristics to facilitate monitoring of catchment treatment plans undertaken in the area . (NHPC[4], 1998).

In the light of site specific problems , the assessment of spatial run-off/erosion patterns, sediment discharges and landslide proneness conditions would help in diagnosing the pretreatment scenario and analyse catchment characters. This would essentially require spatial identification and quantification of various underlying factors such as vegetation and landcover patterns, greenness and production conditions, their associations with topography . For instance it is necessary that monitoring of ecologically sensitive areas such as high slope agriculture and degraded forests, gully and bank erosion areas etc, is highly required to assess catchment conditions. This would only be possible through analysis of integrated units of drainage - slope - elevation - vegetation complexes (Fig 2.2). Accordingly, the study was designed on following lines:-

(a) To provide baseline characterisation of vegetation and landcover, topography conditions, sediment discharges at watershed level.

(b) To assess potential indicators of change which can be used as diagnostic elements for analyzing the changes in the watersheds under treatment.

The baseline assessment at the catchment level was done using only vegetation and land cover conditions and biomass distribution patterns. On the other hand, the detailed analysis of all the five spatial layers was done for the six watersheds which are presently under treatment. These spatial layers are further integrated in GIS domain for identification of sensitive indicator zones.

Assessment of Baseline Vegetation and Landcover Conditions - Rangit Catchment

The false colour composite of satellite data pertaining to Rangit catchment is presented in Satellite Photo plate No 2.1. The multi-spectral remote sensing data which was used in the study enabled to delineate different landcover classes based on the unique spectral properties of each class.

Forests. The vegetation and landcover statistics of the catchment are presented in table 2.1 and the Satellite Photo Plate No 2.1. It is found that dense forest is the predominant landcover in the catchment and constitute 38.96 per cent of the area. These forests are present in extensive patches over high altitudes covering the watershed areas of Reli Chu and Kayam Chu rivers in the North-Eastern part of catchment and Chokchurang Chu and Chhiyjyum khola nallas in South-West part of the catchment. On the other hand in the lower altitude reaches of the catchment , the dense forests are found in fragmented pockets especially in the watersheds which are presently under treatment activity. The open forests constitute only 1.37 per cent of the total catchment and are located in the lower altitudes. The scrub land and degraded forests constituting 2.78 per cent of the catchment are mainly confined to high hill slopes and hill plateaus.

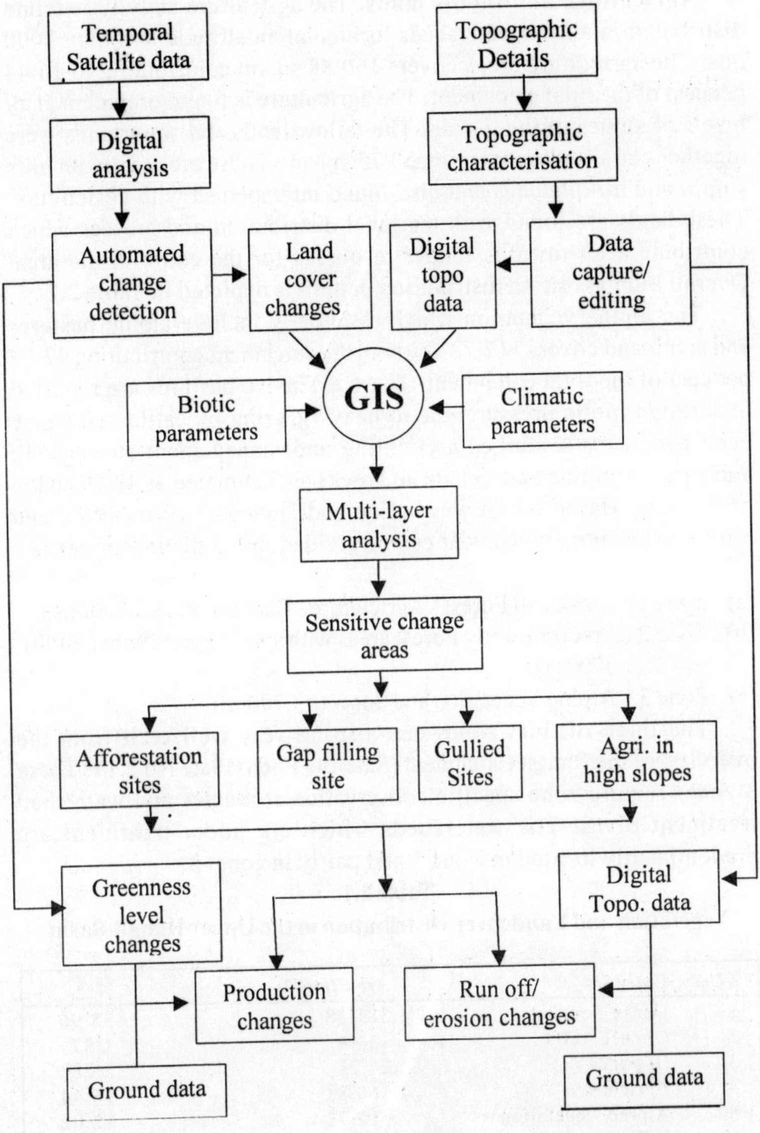

Fig. 2.2 Methodology for Catchment Characterisation

Agriculture and fallow lands. The agriculture activity is found distributed in all the watersheds located at an altitude less than 3000 mts. The agriculture areas covers 160.88 sq km contributing to 16.44 per cent of the total catchment. The agriculture is found prevalent at all levels of slopes of the terrain. The fallow lands and waterlands were together classified as covering 50.75 sq km . These are mostly on high slopes and hill plateaus and also found interspersed with agriculture. These lands are found with marginal distribution and grasses which contribute a lot towards fodder resources for the cattle in the area. Overall biomass strata distribution pattern is depicted in Table 2.2 .

The alpine vegetation which essentially include alpine pastures and scrubland covers 117.72 sq km in the catchment contributing 12.02 per cent of the total catchment. These extensive pastures are reported under high biotic pressure due to heavy grazing by cattle and hence need necessary resources accounting and management strategy. In addition, the alpine barren land and rocks are estimated as 48.79 sq km of the area. Based on the vegetation and landcover distribution and altitude, the entire catchment can be divided into 3 distinct zones :-

(a) Zone 1 Mosaic of Forest - Agriculture - Barren Lands (<2400mt).
(b) Zone 2 Predominantly Forest areas with sparse agriculture (2400-3700mt).
(c) Zone 3 Alpine Vegetation and Snow (>3,700mt)

The three distinct zones can also be very well seen from the overview of the Rangit Catchment (Satellite Photo Plate No 2.1) . These 3 zones require zone specific conservation strategies and watershed treatment plans. The watersheds which are under treatment are predominantly located in Zone 1 and partly in zone -3.

Table 2.1

Vegetation and Landcover Distribution in the Upper Rangit Basin

SL #	*Category*	*Area (km 2)*	*%*
1.	Dense Forests	381.38	38.96
2.	Open Forests	13.44	1.37
3.	Degraded	27.27	2.78
4.	Agriculture	160.88	16.44
5.	Alpine Vegetation	117.72	12.02
6.	Fallow/Wasteland	50.75	5.19
7.	Snow	178.79	18.26
8.	Alpine Rock	48.79	4.98
	Total	979.02	100.00

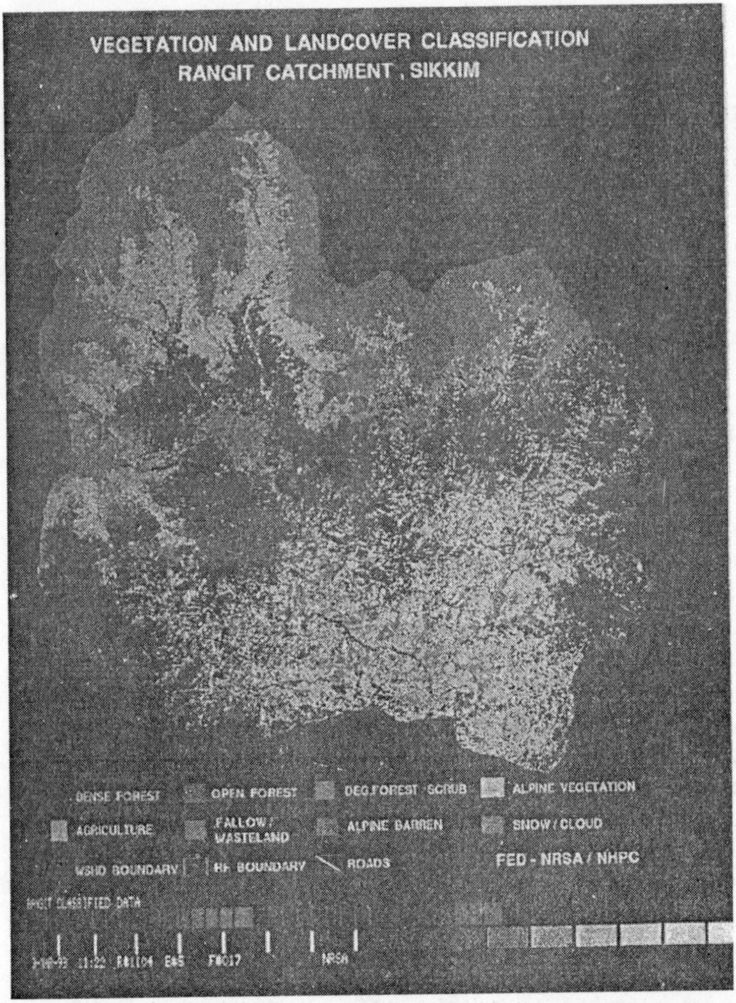
VEGETATION AND LANDCOVER CLASSIFICATION
RANGIT CATCHMENT, SIKKIM
DENSE FOREST
OPEN FOREST
DEG.FOREST SCRUB
ALPINE VEGETATION
AGRICULTURE
FALLOW / WASTELAND
ALPINE BARREN
SNOW / CLOUD
WSHD BOUNDARY
ROADS
FED - NRSA / NHPC

Table 2.2
Biomass Strata Distribution Pattern in Catchment Area

SL #	*Biomass Strata*	*Area (Km2)*	*% of total Area*
1	Highest	78.76	8.04
2	Very high	220.27	22.49
3	High	230.64	23.55
4	Low	203.66	20.08
5	Very Low	245.69	25.09
	Total	979.02	100.00

SECTION-III
HUMAN GEOGRAPHICAL MOSAIC OF SIKKIM

Physical Features

Sikkim extends from 27 to 28 degrees N latitude and 88 to 89 degrees E longitude. Situated in the Eastern Himalayas, its geographical location is invaluable to India. A cursory glance at the location of Sikkim on the map of India reveals the extraordinary strategic importance of the state. Sikkim shares international boundaries with three neighbouring countries, viz China (Tibet), Nepal and Bhutan. Although barely 114 Km long and 64 Km wide, this tiny mountain state merits much attention. (Maitreyi Choudhury[5], 1998).

Having an area of 7096 Sq Km , the whole state is divided into four administrative zones : North ,South, East and West districts with their headquarters at Mangan , Namchi, Gangtok and Gyalshing respectively. As a part of the inner ranges of the Himalayan Mountains, it is wholly a hilly cum mountainous state with varied elevations. The land area can be grouped as :-

(a) *Lower Hills*. Altitude from 300 to 1500 m above mean sea level (M.S.L).
(b) *Mid Hills*. Altitude from 1500 m to 2000 m above M.S.L
(c) *Higher Hills*. Altitude from 2000 m to 3500m above M.S.L.
(d) *Alpine Zone*. Altitude above 3500m above M.S.L to snow line at approximately 5000 m above MSL.
(e) *Snow land*. Higher reaches of mountains and glaciated areas.

Owing to such variations in elevation, the climate of Sikkim varies from sub-tropical to alpine. Accordingly the vegetation can be classified into four major types viz sub-tropical wet type, wet temperate type, mixed coniferous type and alpine scrub and pastures. (NHPC[6], 1998).

About two thirds of the state consists of very lofty mountains which remain perpetually snow covered. Mount Khanchendzonga lies in the North Western side of Sikkim bordering Nepal. Following are the important peaks and passes located in Sikkim (adopted from Maitreyi Choudhury[7], 1998) as given at Table No 2.3:-

Table 2.3

Sectors	*Peaks (above 6000m)*	*passes*
Northern Sector	(i) Lhonak (ii) Sentinel (iii) Chorten Nyima (iv) Khora Khang (v) Khora Tso Gna	(i) Chorten Nyima La (ii) Naku La (iii) Kongra La (iv) Task La (v) Bam Tso La
Eastern Sector	(vi) Pauhunri	(vi) Say Sayla (vii) Ghora La (viii) Khanchung La (ix) Pata La (x) Thangkar La (xi) Nathu La (xii) Jelep La (xiii) Batang La (xiv) Dongkiya La (xv) Jonsang La
Western Sector	(vii) Jonsang (viii) Langbu Pyramid Peak (ix) Tent Peak (x) Nepal Peak (xi) Khanchendzonga (xii) Kabru (xiii) Talung (xiv) Rathong	(xvi) Khsng La (xvii) Daiu La

There is only one major river system in the state of Sikkim that is river Tista. The Chola/Dongkiya Range in the East and the Singalila Range in the West determine the drainage basin of the Tista. The entire state is drained by the Tista, its numerous tributaries, and innumerable sub-tributaries. The Tista river system within the state of Sikkim is listed below at Table 2.4 (Maitreyi Choudhury[8], 1998) :-

Table 2.4
Tista Tributaries

S.No	*Eastern Tributaries*	*Western Tributaries*
(i)	Chhombo chhu	Zemlu (Lachen) Chhu
(ii)	Lhasa Chhu	Rangyong Chhu
(iii)	Kelep Chhu	Rangphap Chhu
(iv)	Gyamthang Chhu	The Great Rangit
(v)	Burum Chhu	
(vi) Gey Chhu		
(vii)	Tarum Chhu	
(viii)	Rasbom Chhu	
(ix)	Lachung Chhu	
(x)	Ong Chhu	
(xi)	Chakung Chhu	
(xii)	Dik Chhu	
(xiii)	Rongni Chhu	
(xiv)	Rangpo Chhu	

The Rangit Chhu, a right bank tributary of the Tista originates from comparatively low altitude area in the south district of Sikkim. However, its chief feeder, the Rathong chhu, originates from the Rathong glacier in the West district. The Rangit receives a large number of tributaries on the way to Tista. Apart from the Rathong Chhu, it is fed by the Rimbi, the Kalej, the Roathak , the Rammam and the Manpur Khola. It also receives some tributaries from the Darjeeling hills of West Bengal, namely, the Little Rangit, the Jhepi and the Ragnu Khola where it is aptly called the Great Rangit . The Great Rangit confluences with the main stream, the Tista at Melli confluence.

The present landscape of Sikkim is in a state of rapid evolution. The ongoing glacial, periglacial, glaciafluvial, fluvial and pluvial activities are continually reshaping the face of this young mountain topography. While the swift flowing rivers are transporting loads of eroded materials by chiselling the valley sideslopes, many of the lakes are choking with sediments. In recent years, there is a noticeable reduction in the size of some of the larger lakes of Sikkim. (Maitreyi Choudhury[9], 1998).

History

The history of the emergence of the kingdom of Sikkim can be traced back to 1642, when Phuntshog Namgyal was consecrated as the first king of Sikkim by three monks at Yoksom in Western Sikkim. These three monks Lhatsum Chhenpo, Nga-Dag Lama and Kathog Lama conferred on him the title of Dharma-Raja (Chogyal). The Dalai Lama recognised the new king as the ruler of Sikkim and sent him a silk scarf, the mitre of Guru Rinpoche and the most precious sand image of the Guru as a 'present'.

Sikkim's boundary, at that time, extended as far as Limbuana in the West, Chumbi valley and parts of Western Bhutan in the East. Southward it extended up to Titaliya and included the whole of Darjeeling district. However, from the beginning of the seventeenth century, the country had to face numerous invasions from its neighbouring countries, Bhutan and Nepal. Consequently, a large portion of its land was annexed by these countries. The Nepalese conquered the lower Tista, Bhutan annexed the entire region on the East of the river Tista including the present day Kalimpong sub - division and Tibet took over Chumbi valley, and thus made the Chola and Jalepa ranges Northern and Eastern boundary (K.S.Singh[10], 1993).

The Anglo-Gorkha war and the signing of the Treaty of Sugauli (1815) between Nepal and the East India Company and signing of the Treaty of Titaliya (1816) between the East India Company and Sikkim introduced a new phase in the history of Sikkim.

By the Treaty of Sugauli the Raja of Nepal agreed never to molest or disturb the Raja of Sikkim, and in case of any difference between the state of Nepal and the Raja of Sikkim, the difference would be referred to the arbitration of the British Government. Under this treaty, Nepal surrendered to the Company, all the hilly regions situated to the East of the Mechi river and to the West of Tista river, which was originally a part of Sikkim but had been annexed by Nepal in 1788-90. By the Treaty of Titaliya the East India Company restored to Sikkim the above. Besides the British secured the right of arbitration in case of dispute between Sikkim and Nepal or any other neighbouring state.

The Treaty of Titaliya marked the beginning of the British influence in Sikkim. By the Treaty of Tumlong (1861), the British could gain a foothold in Sikkim. The British activities in Sikkim gained momentum with the establishment of its Residency at Gangtok and the appointment to the first political officer, JC White, in 1889. The British protectorateship over Sikkim was recognized by China in 1890.

Under the British protectorateship the power of the Maharaja of Sikkim was considerably reduced and he had no role in the administration of the country. It was during the reign of the Maharaja Tashi Namgyal (1914-68) that full power was restored.

The period following the departure of the British from India was marked with the radical change in the political scene of Sikkim. On the one hand independent India signed a standstill agreement with Sikkim in 1948, in order to retain their respective rights and obligations evolved through earlier treaties and conventions; on the other, the educated elites of the state, while drawing inspiration from the freedom movement in India, started various political parties. The main political parties formed in the late forties were Sikkim Praja Sammelan, Praja Mandal and Praja Sudharak Samiti. In December 1947, these parties merged to form a new political party, namely the Sikkim State Congress. The Sikkim State Congress made a three fold demand on the Government; the abolition of landlordism ; the establishment of a responsible government and ; the accession of Sikkim to India. To press its demands for reforms the Sikkim State Congress, besides organising mass demonstrations, started a 'No Rent' campaign, leading no to a political impasse. Consequently, the Indian Government took over the administration and appointed JS Lall as the Dewan of Sikkim. During his term as the Dewan, house tax and forced labour were abolished in order to facilitate the withdrawal of the 'No Rent' campaign.

The Indo-Sikkim Treaty (1950) was signed to give definite shape to Indo-Sikkimese relations. The 1950 treaty was an important landmark in the political development of Sikkim, for with this the government of India revoked all the previous treaties and affirmed the protectorate status of the state under independent India.

Defence, external affairs, customs and communications of Sikkim became India's responsibility. The 1950 treaty affirmed India's authority over Sikkim.

The period between 1950 and 1973 was marked by several important events in Sikkim, like the emergence of various political parties, formation of the state Assembly, introduction of franchise on the basis of party formula. The tension that developed between the Durbar and the various political parties led to the agitation of 1973, the signing of the May Agreement of 1973, formation of a popular ministry in 1974, and finally its merger with the Indian Union in 1975. Sikkim became the twenty-second state of the Indian Union on 26 April 1975.

People

Demography

Sikkim has three main ethnic groups; the Lepchas who were the original inhabitants of the state; the Bhutias who migrated from Tibet in the seventeenth century; and the Nepalese whose migration from Nepal started in the nineteenth century.

The first official census of Sikkim, taken in 1891, presents the following figures for different groups of population (Risley[11], 1894) :-

Race/Caste	*Total*
Lepcha	5,762
Bhutia	4,894
Limbu	3,356
Gurung	2,921
Murmi	2,867
Rai/Jimdar etc	2,020
Khambu	1,963
Kami	1,670
Brahman	1,414
Mangar	901
Chhettri	829
Newar	727
Darzi	287

After 1891 only Lepchas, Bhutias and Limbus are taken as distinct ethnic groups while all other groups mentioned above came to be regarded as different castes of the Nepali Society.

Writing in 1909, JC White gave the number of Lepchas inhabiting Sikkim as about 6,000, Bhutias as a little over 6,000 and Nepalease as nearly 50,000 (White[12], 1909).

According to the 1931 Census Reports, out of a total population of 109,808 there were 13,060 Lepchas, 11,955 Bhutias and the rest were mainly Nepalese. The 1971 Census Report gave the percentage of population by language, according to which speakers of the Nepali language constituted 63.97 per cent, the Lepchas and Bhutia languages were spoken by 10.63 per cent and 11.23 per cent of the total population respectively.

Within a decade, there was a significant rise in the population of Sikkim. This was mainly due to the influx of population from other parts of the country after merger of Sikkim with India. Thus, in 1981 the total population of Sikkim was 3,16,385 and in 1991 it rose to 4,06,457. The 1981 Census Report provided the categorisation of population on the basis of 'language mainly spoken in the household' with Bhutia spoken by 6.8 per cent, Lepcha by 7 per cent, Limbu by 5 per cent, Sherpa by 3.4 per cent, Tibetan by 1.34 per cent and Nepali by 61 per cent of the total population (Census Report Sikkim[13], 1981, 1991).

In terms of density of population the East district is the most populated with 43.90 per cent of the population residing in this district, followed by the South District (24.26 per cent), the West District (24.15 per cent) and the North District (7.69 per cent). Although the North District represents 59.56 per cent of the total area of Sikkim, two-third of the region is covered either with perpetual snow or thick forest. It is the most thinly populated district of the state, with only 7 persons per sq. km as against 187 persons per sq.km in the East District.

Districtwise general information on (1991)[14] census is given at Table 2.5

Table 2.5
Districtwise General Information : 1991 Census

Sl No	Item	Unit	North	East	South	West	State
1	District	No	1	1	1	1	4
2	Sub-division	No	2	2	2	2	8
3	Zila Panchayat Wards	No	20	27	23	22	92
4	Gram Panchayat Wards	No	98	269	243	263	873
5	Gram Panchayat Units	No	20	48	42	49	159
6	Revenue Block (Vill)	No	53	134	145	121	453
7	Towns	No	1	3	2	2	8
8	Area	Sq.km	4226	954	750	1166	7096
		%	59.56	13.44	10.57	16.43	100
9	Density	Per sq km	7	187	131	84	57
10	Population	Nos	31240	148452	98604	98161	406457
		%	7.69	43.90	24.26	24.15	100
11	Urban Population	%	20	7.84	63	43	9.10
12	Rural population	%	7.49	36.06	23.63	23.72	90.90
13	Scheduled Tribe Population	No	17301	37632	16671	19297	90901
		%	19.03	41.40	18.34	21.23	22.36
14	Scheduled caste Population	No	1111	12482	5564	4927	24084
		%	4.61	51.83	23.10	20.46	5.93
15	Decennial Population Growth	1981-1991	18.09	28.60	29.20	31.14	28.47
16	Sex Ratio	F/000 Males	828	859	892	915	878
17	Literacy	%	53.55	65.13	54.08	45.62	56.94
18	Households	No	6658	34241	17924	17506	76329
		%	8.72	44.86	23.48	22.93	100
19	Total Main Workers	%	42.9	37.14	41.8	44.32	40.45
20	Marginal workers	%	1.23	1.37	0.98	0.54	1.06
21	Non-Workers	%	55.87	61.49	57.22	55.14	58.49

Dress

Lepcha. The traditional dress of the Lepcha men is called *gada*. It consists of a long, beautiful striped cloth loosely worn down to the knee leaving one or both arms free. The trousers which reaches up to the calf only are called the *phomu*. Women wear the *dumdyan*, a full length robe, held at the shoulders by a pair of brooches and fastened at the waist by girdle of silver chains or simple hand-woven strip of cloth, known as the *namrek*. Over the *dumdyan* a loose coat called the pago is worn. A scarf (*thorough*) is tied round the head. Shoes for both men and women are called *hlom* (pronounced lhom)(Sanchita Ghatak[15], 1993).

Bhutia. The Bhutia costume for both sexes is called the *kho*. The *kho* worn by the male is called the *po-kho*, whereas that worn by a woman is called the *mo-kho*. Underneath the *kho*, a man wears a double breasted, high necked shirt, the *tson-ta-ti*, and pants (*gya-ton*). A woman ears a long, loose-sleeved blouse (*honju*) underneath the *kho*. A married woman ties a striped apron (*pangden*) over the *kho* at the waist. The quality of the clothes worn both by men and women vary according to the economic status of the person ranging from simple cotton to Chinese brocade and silk. The *kho* is held at the waist by a long piece of cloth (*kera*).

Nepali. The traditional costume of Nepali men consists of the *daura* and the *suruwal*. The *daura* is a knee-length double-breasted shirt, fastened at four points across the body by a string of cloth called the *tuna*; the trousers is called the *suruwal* . A seeveless coat over the daura and a cap called the *birkhe topi* complete a Nepali man's dress. Nepali women wear the *choubandi*, a dauble-breasted blouse held at four points across the body by the *tuna* and the *phariya* or the *gunui*, which is worn at the waist and reaches down to the ankle with tiny pleats in front. It is held at the waist by a long strip of cotton cloth (*patuka*). They cover the head with a triangular or rectangular piece of cloth (*majetro*) and tie a cloth around the bust down to the hips (*hembari*).

Jewellery

Lepcha. Lepcha men do not wear any jewellery except amulets given by the monks. The women are very fond of wearing colourful necklaces made of turquoise, coral, amber or simple glass beads (*lyak*), ear rings (*nay-kong*), bangles (*kaghu*) or rings (*kakyup*).

Bhutia. Bhutia women were necklaces of various designs called the *khou* the *ka-chung*, the *zuri*. The *chyap-chyap* is worn with the *khou* as a *brooch*. The Bhutia names for different items of ornaments are the *len-dab* (ear ring), the *zuku* (ring), the *deu* (brecelet). All the ornaments are made of gold, studded with various precious stones like *yeu*, *dhow* and *zi*. The zi is considered to be the most sacred and precious stone (SRIT Bulletin[16], 9189).

Nepali. Nepali women wear mostly gold and silver ornaments. The traditional ornaments are ear rings (*chyaptesun, godavari*), necklaces (*tilhari kantha*), and bracelets (*chura*), nose rings (*mundri* , *dungri bulaki*) and anklets (*kalli*). *Kallis* are made of silver. The different names for the ornaments represent the different designs of the items. At the time of marriage the bridegroom ties around the neck of the bride strings of green tiny beads (*potey*), that serve for a sign for the married Nepali woman. Nepali men do not wear any jewellery except the ring (*aunthi*) (Subba[17], 1989).

Agricultural Implements

In Sikkim people are yet to make use of improved and mechanized implements in agriculture mainly because of the nature of the terrain. All the three communities use identical implements like the wooden plough, commonly known as the *halo*, which is widely used. The *kanta*, a spade-like tool with sharp teeth, is used for digging the soil. Other tools widely used include the sickle, called the *hasiyan* by the Nepalis and the *nga-tso* by the Bhutias; the sickle with a row of sharp teeth called the *kachiyan* by the Nepalis and the *nga-ra-tso* by the Bhutias, the bullock-driven ground-leveller, called the *dante* by the Nepalis and *chhyi-ri* by the Bhutias. For cutting wood and splitting bamboos or for such other work, Nepalis use the *khukri* whereas the Bhutias and the Lepchas use the *bentok* and the *bamphok* respectively (K.S. Singh[18], 1993).

Food

The staple food of the Lepchas is rice. These days they sometimes take *chapatis* too. They also take roasted or boiled maize. They are non-vegetarian and eat all kinds of meat, fish and birds. It is said that earlier they would take the meat of monkeys and snakes. The present generation tends to keep away from these. They eat various types of tubers and vegetables. One of the most important items in the Lepcha diet is *chi*, home-brewed millet alcohol. A Lepcha who does not drink the *chi* is hard to find. (Foning[19], 1987).

The traditional staple food of the Bhutias includes the *tsampa*, the *thukpa* and the *momo*. These days they also take rice and sometimes *chapatis*. They are non-vegetarian and eat egg, fish, chicken, mutton, pork and beef and sometimes buffalo meat. Dried beef is a delicacy for them. They also eat various types of tubers and vegetables. During festivals they make several kinds of sweets and salted refreshments called the *khabse* and the *fongui amcho*. They are very fond of drinking the home-brewed alcohol called *chhang* (called *chi* by the Lepchas and *janr* or *rakshi* by the Nepalese)(Sanchita Ghatak[20], 1993).

The staple diet of the Nepalis is rice and pulses and sometimes *chapattis*, or the oil-fried small chapatis (*puri*). The various groups of Nepalis have different dietary pattern with regard to meat and drinking alcohol. The Bahuns or Brahmans are mostly vegetarian and do not drink alcohol; instead they eat different type of tubers, vegetables and are very fond of drinking milk and curd. People of the Chhettri, Thakuri and Majhi groups are non-vegetarian but eat only egg, fish, chicken and mutton, and do not eat pork and beef. The Yakhas, Damais, Mangars, Gurungs, Kamis,Sarkis and Bhujels take beef and buffalo. The Newars eat egg, fish chicken, mutton and buffalo but do not eat pork and beef whereas the Kagates, Limbus Sherpas, Tamangs take egg, fish, chicken, mutton, pork, beef and sometimes buffalo. Some of the sub-castes among the Tamangs and Gurungs do not eat pork or beef. All the communities mentioned above take tubers and vegetables which are either locally grown or bought at the market. Except the Brahmans, all other groups among the Nepalese take home-brewed millet or rice alcohol (*janr, rakshi*). Among the Nepali Hindus, there are followers of various sects like Kabir Panthis, Pranamis, Raidas. These people are strictly vegetarian and some do not take even garlic and onion.

The festival delicacies of the Nepalis include *chel roti, phuraulo*. On the Maghe Sankranti day, various kind of boiled tubers and roots and *phuraulo* are taken, and *chel roti* is made especially during Tihar or for a marriage or any other ceremony except the death ceremony. (Subba[21], 1989).

JP Tamang[22] (1998) explains that more than seventeen varieties of indigenous fermented foods are traditionally prepared and consumed by people at house hold level for centuries in Sikkim as shown in Table 2.6

Table 2.6

Indigenous Fermented Foods of the Sikkim Himalaya

Fermented food	*Substrate*	*Nature and use*
Common fermented foods		
Kinema	Soybean	Cooked soybeans showing stickyness with Typical flavour, side dish
Gundruk leafy vegetable		Dried, sour-acidic taste; soup, pickle
Sinki	Radish root	-do-
Selroti	Rice/wheat flour	Pretzel-like, deep fried, confectionary bread
Dahi	Milk	Thick gel; savory
Mohi	Milk	Butter-milk; non-alcoholic drink
Chhurpi	Milk	Soft-mass is used for curry, pickle Hard mass is consumed as masticatory.
Sukako masu	Mutton/beef/ pork/yak	Smoked meat; curry
Common fermented beverages		
Marchaa Rice,wild herbs, spices		Solid, dry, white cake-like starter culture to Forment beverage
Kodo Jaanr	Finger millet	Midly alcoholic, slightly acidic-sweet Beverage
Bhaate Jaanr	Rice	-do-
Makai Jaanr	Maize	-do-
Raksi	Starchy substrate	Distilled alcoholic drink
Lesser-known fermented foods		
Masauyra	Black gram	Ball-like hollow; spicy condiment
Mesu	Bamboo shoot	Sour-acidic flavour; pickle
Khalpi	Cucumber	-do-
Somar	Milk	Soft-mass paste, brownish colour with flavour condiment

Religion

Religious Groups

Two important religions professed by the people of Sikkim are Mahayana Buddhism and Hinduism.

The introduction of Buddhism in Sikkim coincides with the establishment of a political institution in Sikkim in the seventeenth century. The three monks who had come to Sikkim from Tibet started propagating the religion and constructing monasteries. Mahayana Buddhism was the state religion till the abolition of the system of hereditary monarchy in 1975.

The two sects of Mahayana Buddhism in which the majority of the Buddhists in Sikkim are organized are the Nyingmapa and the Karma-Kadgupa.

A majority of Bhutias and the Lepchas are Buddhists. There are some Buddhists among the Nepalese like the Tamangs and Sherpas, but the majority of them are Hindus, worshipping various Hindu gods and goddesses. One of the striking features of the religious practices of the Nepalese is the worship of the family diety (Kul-Devata). Those who worship the Kul-Devata have a separate altar to which only close family relations have access. Generally the rites and rituals are performed by priests of the same caste. (Subba[23], 1989).

Besides of Buddhists and Hindus, there are a few Christians, Muslims, Sikhs and Jains. The classification of population on the basis of religion (1981 Census) is as follows :-

Religion	*Percentage to Total Population*
Buddhist	28.71
Hindu	67.25
Christian	2.22
Muslim	1.03
Sikh	0.10
Jain	0.03

Religious Institutions

Monasteries. There are about seventy seven monasteries in Sikkim. The oldest monastery at Dubdi in Western Sikkim was constructed soon after the consecration of Phuntshog Namgyal. The monasteries of Tashiding, Rinchangpong, Sangacholling and Pemayangtse were constructed in the latter half of the seventeenth century or in the early eighteenth century. A few important monasteries of the Nyingmapa sect in Sikkim are Pemayangtse, Tashiding, Sangacholling, Enchey and Tsuk-la-Khang. (D.P. Boot[24], 1996).

The first monastery of Karma-Kadgupa was constructed in 1730 at Ralang in South Sikkim by the fourth Chogyal of Sikkim, Gyurmed Namgyal, in pursuance of a promise given by him to the twelfth incarnation of Gyalwa Karmapa at Tsurphung monastery in Tibet. Other important monasteries of this sect are at Phodong and Rumtek. Rumtek is the present seat of the current 17th Karmapa. (RR Dhamala[25], 1993).

Temples. The Thakurbari at Gangtok was established in 1935. Besides this there are about fifty temples scattered all over the state. There are four churches at Gangtok and six others in other parts of Sikkim. The lone mosque in Sikkim is located at Gangtok. (K.S.Singh[26], 1993).

Languages

The three major languages spoken in Sikkim are Nepali, Bhutia and Lepcha. Due to the overwhelming presence of the Nepalese, Nepali has since long become the *lingua franca* of the state.

The Bhutia and Lepcha languages belong to the Tibeto-Burman group, while Nepali is an Indo-Aryan language.

There are four official languages of the state, viz Lepcha, Bhutia, Nepali and Limbu, which are taught in the schools also.

The 1981 Census gave the following figures on the basis of language spoken in the household:-

Bhutia	*Lepcha*	*Limbu*	*Nepali*	*Sherpa*
21,548	22,391	17,922	192,891	10,726

Besides these languages, the several dialects prevalent within the different Nepali castes include Rai, Tamang, Gurung, Mangar, Newar etc. However, these dialects are very rarely spoken even within the same community. Speakers of these dialects usually speak the standard Nepali.

There are also speakers of many other languages like Hindi, English, Punjabi, Bengali, Tamil, Telugu, Malayalam, Manipuri and Urdu.

Settlement Patterns

Ever since Sikkim was exposed to outside influence it has been the meeting place of two diverse cultures brought in by the two immigrant communities, viz. The Bhutias and the Nepalese. While the Bhutias brought with them a part of Tibetan cultural life including the language, religion and an economic system which was a combination of pastoralism and semi-settled agriculture, the Nepalese brought with them a part of Nepal's cultural life including the Nepali language, Hinduism and settled cultivation especially terraced cultivation.

While the cultural influence of the Bhutias on the Lepchas, especially the conversion of the Lepchas into the fold of Buddhism, prevailed, the Nepalese retained a distinct identity. Initially, the Bhutias coming from the cold and dry climate of Tibet were resistant to the warmer regions. As a result, their habitation came to be concentrated in the higher regions. Consequently the Lepchas retreated to the remote valleys and forest-clad mountains. The Nepalese began to occupy the warmer regions. The Limbus were mostly scattered in the Western part of Sikkim. In course of time, however, the settlement pattern became more dispersed. At present, all communities are scattered over all parts of Sikkim. The two pockets of Sikkim where one can find the dominance of one ethnic group are Dzongu and the two valleys of Lachen and Lachung of North Sikkim. (Sengupta[27], 1985).

The Dzongu area is situated on the Western side of the river Tista in North Sikkim. It consists of thirteen revenue blocks. In order to preserve and protect the Lepcha culture, it was declared a Lepcha reserve. There is a legal restriction on other communities trying to settle in Dzongu. Still, barring the Sakyong and Pentung blocks, all the other blocks have some proportion of households belonging to other communities. The total population of Dzongu 1981 Census) was 7745, of which 56 per cent are Lepchas. The Lepchas are primarily cultivators, with 95 per cent of the households engaged in agriculture.

The two valleys of Lachen and Lachung in Northern Sikkim are inhabitated by Bhutias, who call themselves People of Lachen (*Lachenpa*) and people of Lachung (*Lachungpa*). According to the 1981 Census, there were 234 households with a population of 1508.

The people of these valleys are essentially herdsmen and traders. Till the closure of the border with Tibet, they had been mainly engaged in trans-border trade. The people are now mostly engaged in agriculture including horticulture.

House pattern

Most of the houses of the Bhutias and the Lepchas are two-storeyed rectangular structures. The ground floor is partially enclosed where domestic animals especially cattle are kept. The houses are made of bamboo splits plastered with mud. The roof is thatched with reed, straw and bamboo. The Nepalese usually build their house right on the ground as they keep their domestic animals in a different shelter. They also use bamboo splits and mud. Some of the houses are also made in wooden planks. Presently corrugated tin roofs are replacing the thatched roofs of the Bhutia, Lepcha and Nepali houses. In most of the urban and semi-urban areas R.C.C. buildings are a common sight. Most of the houses consist of two or more than two rooms depending upon the economic condition of the family. In every traditional house there is a separate room for an altar which also serves as store room for the grains. (Bhasin[28], 1995).

Major Festivals of Sikkim

The festivals of Sikkim can be said to belong to two cultural streams based on religion, the Nepali culture and the Bhutia-Lepcha culture.

Two important festivals of the Nepalese, the Dasain and the Tihar, are known as the Dussehra and Diwali respectively in other parts of India.

Dasain is celebrated during the months of September-October (Ashwin in the Nepali calendar). During Dasasin, the goddess Durga is worshipped, whereas Vijaya Dashami is celebrated with much merry-making. On this day, elders of the family i.e., the grand-parents, parents and other, put *tika* (a mixture of curd and rice, known as *achata* and vermilion) on the foreheads of the younger people with their blessings. Coloured *tika* is not put on the forehead of a widow, instead it should be of white colour. The *tika* celebration continues for five more days. On the first day is the *Kag Tihar* which literally means "the festival of the crow (*kag*). On this day delicacies are prepared and given to the crow. The second day is the *Kukoor Tihar*. On this, special attention is given to the dogs. Flower garlands are put around the necks of dogs who are given special food, consisting of meat and cooked rice. The third day is the *Gai Tihar*. On this day cows are worshipped and given special food. On this evening each household worships the Goddess Lakshmi. The houses are decorated with marigold garlands and rows of lights are lit in the evening. After this, groups of girls visit the houses and sing the special song know as the *bhailo*. Such singing groups are called the *bhailinis*. The next day is the celebration of *Goru*(*ox*) *Tihar*, and on the following day is the *Bhai-tika*. On this day, a sister (*cheli*) puts a multicoloured *tika*, usually of three or four colours, on the forehead of her brother (*maiti*). On the two days of *Goru Tihar* and *Bhai-tika* groups of boys visit houses singing traditional songs (*dewsi*) normally accompanied on the drum (*madal*). (Subba[29], 1989).

Another important festival of the Nepalese is the Maghe Sankranti which falls usually in the middle of January. On this occasion prayers are offered at the confluence of rivers and people take a dip in the river. The special food items taken on the day comprise various types of boiled roots and tubers and *phuraulo*.

All the Hindu Nepalese irrespective of caste observe these festivals. Interestingly, the Tamangs, who are Buddhist Nepalese, observe these festivals with equal fervour.

The important festivals of the Buddhists of Sikkim are the Tibetan New Year, the Sagadawa, the Pang-Lhabsol, the Losoong, and the Lhabab-Duchen.

The Tibetan New Year which is known as the Lossar marks the preaching of Dharma by Lord Buddha to the devil, and is usually celebrated in the month of February-March for seven days with much merry-making.

Sagadawa is an auspicious month for the Sikkimese Buddhists. Prayers are held throughout the month in various monasteries. On the full moon of the fourth month of the Tibetan calendar is the main celebration. This religious festival is considered the most sacred , because it signifies the birth of Lord Buddha , his enlightenment and attainment of *Nirvana*. A colourful procession of monks with geylings and rabdung go around the major thoroughfares of the town carrying the Kangyur texts on their heads.

Pang-Lhabsol is a festival to worship Mt. Khanchendzonga, the guardian deity of Sikkim. It is held on the fifteenth day of the seventh month of the Tibetan calendar.

Losoong, called Namsoong by the Lepchas, is also known as Sonam Losoong. It is essentially the festival of the harvest. On this day, newly harvested paddy and fruits are offered to the gods. It is celebrated on the first four days of the eleventh month of the Tibetan calendar.

Lhabab-Duchen is another important festival of the Buddhists of Sikkim, celebrating the descent of Lord Buddha from the thirty-third heaven. (SRIT Bulletin No. 3[30], 1993).

Socio-Political Institutions

Social Organisation

The social organisation of the Lepchas, Bhutias and Nepalese present three distinct patterns. While the kinship and clan association give a Lepcha a sense of social and territorial belonging, this is absolutely absent in the case of the Bhutia. A Bhutia seldom uses his clan name as he does not possess a strict social organization of clan. Many Bhutia clan names are associated with a locality but there is no sense of a genealogical identity involved. They do not regard the clan members as descendants from a common ancestor as in the case of the Lepcha.

In the Nepali society the caste names do not provide any sense of territorial belonging. It is also characterised by caste hierarchy and the practice of untouchability, which however is now on the wane. The chief function of caste is in the regulation of marriage. The Lepchas did not have any social organisation. In the past, the *mandal* who was the collector of revenue used to function as the advisor to the villagers in case of any necessity. The post of *mandal* was hereditary and still continues to be so. However, of late, the constitution of statutory panchayats has considerably minimized the social function of *mandal*.

The *kyi-dug* constitutes the social organisation of the Bhutias. Generally the *kyi-dug* is identified with one territorial clan, like Thomorva, the Ardowa or the Khampa. The main function of the *kyi-dug* is to help its clan members at times of need, for instance, if any death occurs, every family is expected to send a member to help in cremation. If a family fails to do so, the *kyi-dug* would impose a fine on the erring member. Its office-bearers are elected by the members from amongst themselves.

The Lachenpa and Lachungpa of North Sikkim have the *Zumsha* organisation in existence from very early time; JC White[31] (1909) described the *Zumsha* as being 'very unusal and almost a communistic government', where everything from carrying of load for visiting government officers to allotment of land to different members was decided upon by the whole population of the village. The village headman (*pipon*) was elected by adult members drawn from each of the family for a term of two years. Besides the *Zumsha*, there was another body of village elders called the *Gyemmi*, for aiding and advising the *pipon* in functions involving the community as a whole. The functioning of the *Zumsha* in these two villages proved to be so useful that the Government of Sikkim allowed it to function as a rural self-government unit when the Sikkim Panchayat Act 1965, and later the Sikkim Panchayat Act 1982 were enacted.

Among the Nepalese there are various caste organizations generally known as *Samaj*, like the *Tamang Samaj*, the *Rai Samaj*, the *Kami Samaj* etc. These councils function as agencies of social control. Sikkim was under a system of hereditary monarchy till 1975. It was brought under the Bitish protectorateship after the Tumlong Treaty of 1861. The country was administered by the British political officer with the help of a council consisting of influential *Kazis* and representative of the monastery. The full authority of the Maharaja was restored in 1917. (Sengupta[32], 1985).

State Council. The first state council was constituted in 1953. It consisted of a president nominated and appointed by the Maharaja. Twelve elected members of whom six belonged to the Lepcha-Bhutia and six of the Nepali communities, and five members nominated by the Maharaja. In 1958, the number of council members was raised to twenty with the introduction of one general and one reserved (*Sangha*) seat each and an addition of one more nominated member. In 1966 the number was raised to twenty-four with an addition to the number of Bhutia-Lepcha and Nepali seats and the introduction of a scheduled caste seat and a Tsong seat.

Executive Council. According to the Maharaja's proclamation of 23 March 1953 the executive council was constituted consisting of the *Dewan* 'by virtue of the office which he holds under the Maharaja' and 'such numbers of elected members of the state council as may be appointed by the Maharaja from time to time'. The members of the executive council were to hold office 'during the Maharaja's pleasure' and were to be 'responsible to him for the executive and administrative functions of the government'.

The departments placed under the executive council were education public health excise, press and publicity, transport, bazaars, forests and public works.

State Legislative Assembly. Following the agitation of 1973 there was some change in the structure of the political institutions of Sikkim. Under the May agreement (1973) election to the Assembly was to be held under the aegis of the Chief Election Commissioner of India. The new Legislative Assembly consisted of 32 seats including two reserved seats, one each for the scheduled castes and the *Sangha*. This maintained the parity between the two communities.

In 1978, a presidential ordinance laid down a new formula for the distribution of seats in the Assembly, abolishing the party formula; seats were now reserved for the Bhutia, Lepcha, *Sangha*, and scheduled castes. This distribution of seats in the Assembly in now as follows (Sengupta[33], 1985):-

Community	*Number of Seats*
Bhutia - Lepcha	12
Scheduled Castes	2
Sangha	1
General	17
Total	32

Economic Development

Five-Year Plans

By the provisions of the Indo-Sikkim Treaty (1950) the Government of India had taken special responsibility in initiating planned economic development in the state. Accordingly, with financial assistance from India, a Seven-Year plan (1954-1961) was undertaken with a total outlay of Rs 324.00 lakhs, followed by three Five-Year plans from 1961 to 1976.

The total outlay for these plans were Rs 641.00 lakhs for 1961-66 Rs. 972.00 lakhs for 1966-71 and Rs 1875.00 lakhs for 1971-76 plan periods. In these plans special emphasis was laid on the development of infrastructures like road and communication for which about 48 per cent of the total outlay was allocated. Agricultural development did not receive much attention for which only 16 per cent was allocated during the second and fourth plans. In the third plan (1966-71) 44 per cent of the total outlay was allocated to the agriculture sector. (Economic and Statistic Bureau Sikkim[34], 1992-93). In the post-merger period there was a shift in the thrust of the plan, and agricultural development was given priority in the state development plans. During the fifth plan period the agriculture department was able to 'develop a basic infrastructure including establishment of Regional Centers, Regional Sub-Centres, VLW. circles and specialised farms. Attention was also given to the development of a cadre of technically qualified personnel with a view to create the necessary environment for implementing plan programmes in the agricultural sector'.

Environment was given its due in Eighth Five year plan, outlay and expenditure of which is given at Table 2.7 (Stats Bureau Sikkim[35], 1998).

Development Activities.

Agriculture. The Department of Agriculture and Animal Husbandry was established in 1954 with the initiation of the Seven-Year plan. The activities of the department increased considerably after 1975, with notable developments like the establishment of the nine regional centres, five regional sub-centres and establishment of one hundred VLW circles in order to help the farmers adopt developed farming techniques.

Table 2.7
Eighth FYP (1992-1997) : Outlay and Expenditure

Rs. in Lakh

Sl. No.	*Major Head/Minor Head of Development*	*Original Outlay*	*Actual Expend-iture*	*Percentage Distribution*
I.	AGRI & ALLIED SERVICES	7350	12881	16.59
II.	RURAL DEVELOPMENT	1280	2865.97	3.69
III.	IRRIGATION &FLOOD CONTROL	1300	1426.62	1.84
IV.	ENERGY	13450	13407.17	17.27
V.	INDUSTRY & MINERALS	2080	3053.74	3.93
VI.	TRANSPORT	9000	10013.03	12.90
VII.	COMMUNICATIONS	-	-	-
IX.	SCIENCE, TECHNOLOGY & ENVIRONMENT	400	404.68	0.52
X.	GENERAL ECONOMIC SERVICES	1100	1256.34	1.62
XI.	SOCIAL SERVICES	-	-	-
	General Education	5500	8633.32	11.12
	Technical Education	280	12.31	0.02
	Sports and Youth service	280	260.74	0.34
	Art & Culture	330	393.94	0.51
	Medical and public Health	5220	7060.95	9.10
	Water Supply and sanitation	3000	6287.26	8.10
	Housing (inc. Police Housing)	550	3935.99	5.07
	Urban Dev (incl State)	450	1023.66	1.32
	Information and publicity	250	287.46	0.37
	Welfare of SC/ST & Other B Classes	350	680.65	0.88
	Labour & Labour Welfare	30	27.47	0.04
	Social Security & Welfare	150	495.39	0.64
	Women & Child Welfare Nutrition	400	873.29	1.12
	Total : XI	16790	29972.43	38.61
XII	GENERAL SERVICES	1250	2347.13	3.02
	GRAND TOTAL	54000	77628.11	100.00
	Non-Plan Deficit	1000		

Although 80 per cent of the total population is dependent on agriculture, only 15.69 per cent of the total geographical area is put to agricultural use. Area under different land use in Sikkim as per agriculture Census is given in Tables 2.8 to 2.11.

With this acute shortage of cultivable land in Sikkim, the aim of the agriculture department has been to increase the productivity of land, rather than bringing more land under cultivation. Its strategy has been to replace traditional farming with improved techniques like the use of improved seeds, fertilizers, effective plant protection and proper storage of farm produce. Efforts are being made to introduce multiple cropping so that the farmers make best use of their resources like land, man power and natural sources.

The main crops grown in Sikkim are maize , rice , millet , wheat , barley , buckwheat etc. The most important cash crop is cardamom and the cash fruit is orange. The department is given special attention to promote the cultivation of these two items.

Industry. Before 1975 there were only a few industries in Sikkim, the most notable ones being the Food Preservation Factory, the Sikkim Distilleries and the Sikkim Jewels. Before 1974 the industrial activities were looked after by the Trade, Industries and Commerce Department. (Manas Das Gupta[36] , 1992).

Soon after 1975 all the four districts were declared industrially backward and a Directorate of Industry was established in 1976 to take over all responsibility of promotion and development of industries in the state.

Most of the industries that exist in Sikkim are either small scale or medium scale. About seventy-two registered units are functioning in the small scale sector. Special attention has been given to promote this sector of industry.

Industrial development is mainly state-sponsored and an emerging class of entrepreneurs are being given assistance by the government. The establishment of the Sikkim Industrial Development and Investment Corporation in 1977 was step towards this direction.

Table 2.8

Agriculture Area under Different Land Use in Sikkim as per Agriculture Census

Sl. No.	*Classifications*	*Total Area in Hectares* 1976-77	1980-81	1990-91
LAND UNDER OPERATIONAL HOLDINGS				
1.	Net Area sown	64,927 (9.15)	78,321 (11.04)	63,254 (8.91)
2.	Area under Current Fallow	501 (0.07)	4,428 (0.62)	3,906 (0.55)
3.	Other un-cultivated area Excluding fallow land	4,925 (0.69)	4,560 (0.64)	10,830 (1.53)
4.	Fallow otherthan Current fallow	944 (0.13)	9,474 (1.34)	9,204 (1.30)
5.	Culturable West Land	1153 (0.16)	681 (0.10)	9,807 (1.38)
6.	Land not available for Cultivation	6613 (0.93)	11,604 (1.64)	14,300 (2.02)
(A)	TOTAL OPERATED LAND % To total Geog.Area	79062 (11.14)	1,09,068 (15.37)	1,11,302 (15.69)

	LAND UNDER NON-OPERATIONAL HOLDINGS			
1.	Permanent pastures & Grazing Land incl culturable waste	87,188	72,937	1,20,299
2.	Land put to non-agricultural use	69,96	85,362.	
3.	Barren Land	2,04,01	1,80,250	1,60,000
4.	Land under Miscellaneous Trees and Grovers etc	4,17	5,450	2,98,000
5.	Forest	2,85,21	2,56,533	
(B)	TOTAL UN-OPERATED LAND	6,30.538	6,00.532	5,98.298
	Total Geographical Area of Sikkim = (A) + (B)= 709600 hec.			

Table 2.9
Operational Holdings 1990-91 Highlights

Sl. No.	*Particulars*	*Operated Area*		*% to total Geog.*
		Hectare	*%*	*Area*
1.	Net Area Sown	63,254	56.83	8.91
2.	Area under current Fallow	3,906	3.51	0.55
3.	Net Cultivated Area	67,160	60.34	9.46
4.	Total Uncultivated Land	29,841	26.81	4.24
5.	Land not available for Cultivation	14,301	12.85	2.02
	Total Area	1,11,302	100.00	15.69

Table 2.10
Operational Holdings by Type: Area (in Hec)

Sl NO	*Partuculars*	*1980-81*	*1990-91*	*Variation*
1.	Individual Holdings	4120	3027	(–)26.52
2.	Joint Holdings	107095	110022	2.73
3.	Total	1973	1279	(–) 35.17
4.	Institutional Holdings	-	-	-
5.	Total	109068	111302	2.05

Table 2.11
Availability of Land per Head in Sikkim (in Hec.)

Sl NO	*Type of land*	*1971*	*1981*	*1991*
1.	Net Cultivated land	0.31	0.27	0.17
2.	Operated Area for Agricultural use	0.38	0.35	0.28
3.	Land for Non - agricultural use	0.33	0.28	0.21
4.	Pasture & Culturable Wasteland	0.4	0.24	0.18
5.	Forest	1.26	0.83	0.65

One important industry that has been drawing the attention of the government is the tea industry, established in 1965. Recently around 370 acres of land have been converted to tea plantation. In order to give a boost to this industry a programme for decentralised cultivation of tea in private lands has been taken up for implementation. Other newly established industries include the Sikkim Flour Mill, Yoksom Breweries, Soap Industry and Denzong Rubber.

Education. There are at present 1559 schools in Sikkim and 1,36,976 school-going children. The literacy rate among the males stands at 65.7 per cent while 46.76 per cent of the women are literate. (Economic and Statistics Bureau[37], 1999).

Health. Health facilities have been expanded to cover the entire population. There are five hospitals, 24 primary health centres, 147 primary health sub-centres. The rate of infant mortality is rapidly declining, which stands at 46 per thousand births at present. One Referral Centre has recently come up in Gangtok having al modern medical facilities.

Social welfare and Nutrition programmes cover a large segment of the population. A massive programme to ensure availability of drinking water in the rural areas has been implemented and out of 440 villages 168 are fully covered while another 200 villages are partially covered. (Economic Statistics Bureau[38], 1998).

References

1. Maurice F. Strong; *The Himalayan Dilemma,* UN University, New York 1989.
2. Barua , Maitreyi ; *Habitat and Economy in An Eastern Himalayan Enclave: A Case Study of the Rangit Basin* , University of Calcutta, 1988.
3. Forestry and Ecology Division , *Catchment Characteristics Rangit Hydel Power Project* Sikkim, NHPC Faridabad, 1998.
4. Forestry and Ecology Division, *Catchment Characteristics Rangit Hydal Power Project,* NHPC Faridabad, 1998.
5. Maitreyi Choudhury; *Geography of Sikkim* in Sikkim : Perspectives for Planning and Development, Sikkim Science Society, 1998.
6. Forestry and Ecology Division; *Catchment Characteristics Rangit Hydel Power Project,* NHPC Faridabad, 1998.
7. Maitreyi Choudhury; *ibid.*
8. *ibid.*
9. *ibid.*
10. Singh K.S; *People of India* : Sikkim, Anthropological Survey of India, Sea Gull Books, Calcutta, 1993.

11. Risely, H.H; *The Gazetteer of Sikkim*, D.K. Publishers, Delhi; Reprinted 1995, Original Print 1894.
12. White, J.C; *Sikkim and Bhutan*, Manas Publications, Delhi, 1909.
13. Census Report, *Sikkim*, 1981, 1991.
14. Census Report; *ibid.*
15. Ghatak, Sanchita; *Bhutias of Sikkim in People of India*; Sikkim, Seagull Books, Calcutta, 1993.
16. Rechung, J.K; *Buddhist Logic,* in Bulletin of Tibetology, SRIT, Gangtok , 1989.
17. Subba, T.B, *Dynamics of a Hill Society*, Mittal Publications, Delhi, 1989.
18. Singh, K.S; ibid.
19. Foning, A.R; *Lepcha: My Vanishing Tribe*, Sterling Publishers, 1987.
20. Ghatak, Sanchita; *ibid.*
21. Subba , *ibid.*
22. Tamang , J.P, *Indigenous Fermented Foods of Sikkim*, in Sikkim; Perspectives for Planning and Development, Sikkim Science Society, Gangtok, 1998.
23. Subba, T.B, *ibid.*
24. Boot D.P, *Monastries in Sikkim*, CHS , NBU , 1996.
25. Dhamala, R.R; *Monastries of Sikkim with special Reference to Economic Structure*, CHS , NBU , 1996.
26. Singh, K.S.; *ibid.*
27. Sengupta, N.; *State Government and Politics*, Sikkim , Sterling Publishers, New Delhi, 1985.
28. Bhasin, M.K.; *Sikkim Himalayas, Ecology and Development*, Kamla Raj Enterprises, Delhi,1995.
29. Subba, T.B.; *ibid.*
30. Mukerjee, Bandana; *Bhutia Culture in Sikkim*, Bulletin of Tibetology, SRIT, Gangtok, 1995.
31. White, J.C, *ibid.*
32. Sengupta *ibid.*
33. *ibid.*
34. Sikkim Economic Survey, Bureau of Economics and Statistics, Gangtok, Sikkim, 1992-93.
35. Sikkim Statistics, Bureau of Economics and Statistics, Gangtok , Sikkim, 1998.
36. Das Gupta, Manas: *Sikkim : Problems and Prospects of Development*, Indus Publishing Company New Delhi.
37. Subba. Jyotsana ; *Sikkim Statistics*, Bureau of Economics and Statistics , Gangtok , Sikkim, 1999.
38. Health information Bulletin; Statistical Cell, Department of Health, Gangtok, 1998.

31. Sikkim [illegible], 2003.

3

Realms of Spiritual Ecology in Upper Rangit Basin

SECTION I
NATURE, RELIGION, SPIRITUALITY AND HUMAN ECOLOGY

Introduction

Religion is a part of culture. It involves a set of beliefs and practices, a social system through which people seek mental and physical harmony with the powers of the universe, through which we attempt to influence the forces of nature, life and death. As such, religion has been an essential attribute of culture, which cannot be left out from the realms of human ecology. (Jordan and Rowntree[1], 1990). Durkheim[2] (1964) describes God as society divinized. The rule of religion thus are the rules of society, and the object of worship is the society itself.

Some of the principal religion's realms as described by Swanson[3] (1960) are :-

(a) Monotheism is positively related to the presence of a hierarchy of three or more sovereign (decision-making) groups in a society.
(b) Polytheism (belief in multiple superior gods) is correlated with social classes.
(c) Belief in active ancestral spirits is correlated with importance of kinship groups.
(d) Belief in reincarnation is correlated with small and nomadic settlement patterns (inspite of its high frequency in India, a highly evolved group of societies with many large settlements).
(e) Belief in witchcraft is most common in societies with few or no other kinds of social controls.
(f) Supernatural sanctions for interpersonal relations are most frequent in societies with interpersonal differences in wealth.

Nature in Spiritual Traditions

Throughout the centuries humans have had a love-hate relationship with the natural environment. This inter-relationship generally corresponds to (Egri,[4] 1999) :-

(a) We depend on nature for the air, water, food, clothing and shelter.

(b) From the natural environment we obtain the resources to be transformed to meet our material needs.

(c) The natural environment serves as the repository for our waste products.

(d) The aesthetic beauty of nature in all its forms is a source of spiritual and cultural inspiration as well as emotional sustenance.

(e) The powerful forces of nature are also the harbingers of destruction and death.

A less benevolent nature brings the droughts, floods, fires, hurricanes, and earthquakes against which humanity is virtually powerless. There are aspects of nature, living and non-living, with which contact is dangerous for humans. Nature reminds us in innumerable ways of our mortality, that individual lives are but precisely brief journies in time.

The human's dependence on nature, its use and misuse of nature, its fear of nature and its expectations from nature found its way into religion and spiritual realms in almost all human societies. Because human feared nature, it became an accepted necessity to appease the natural forces, some of which took the demonical form. Shamans and '*jhakris*' came into being. Nature in all its forms entered the passages within the religious texts. Religious prayers and hymns were composed and sung in reverence to nature. Alongside however, the misuse of nature, and exploitation of all non-human resources on the planet earth continued. Biotic life and abiotic environment were both targetted for exploitation.

The primary concern of this chapter is on spiritual discourse (with special reference to Upper Rangit Basin of Sikkim) as a means of constructing and reconstructing humankinds' relationship with the natural environment. The ways in which spiritual communities, monasteries and clergy have worked for sustenance of environment is presented in succeeding sections of the chapter. How these lessons have been accessed by individuals and collectives to effect transformational change in ecological consciousness and action will be discussed.

Animism : The Nature Centred Spiritual Tradition

Many different peoples of the world believed and some still do-that behind the immediate physical reality of things lie spirits, that even seemingly dead objects, rocks or earth have a living force within them (Toffler, 1980)[5]. Nature centred spiritualities have existed since before the advent of modern human civilisations. For over 20,000 years animistic spiritual traditions have guided indigenous peoples throughout the world. In this holistic and integrative tradition, "spirituality is not a religion with a fixed set of dogmas, but rather, spirituality pervades and infuses all forms of existence-human, animate and inanimate" (Frost and Egri[6], 1994). It informs the view that all that exists in nature is living and sacred, therefore deserving of respect and care (Halifax[7], 1990). The original spiritual traditions of Limbus (Risley[8], 1894); (Subba[9], 1999) and Lepchas (Foning[10], 1987) fall under this category. Even though these two tribal entities have embraced Hinduism and Buddhism during later periods between 17th to 20th centuries, they have maintained most of their original traditions.

Egri[11] (1999) describes three assumptions which underlie the animist belief system:-

(a) The existence of numerous coexisting and interpenetrating worlds of experience.

(b) The need for holistic balance amongst these worlds.

(c) That change is a continuous transformational process.

The nature-centred spiritual traditions emphasize human beings' physical and spiritual connections with an animate and sacred for respect of all that exists in the natural environment, as well as the need for transformational change to regain holistic balance and harmony in the cosmos. This in a nutshell is also the concept of deep ecology.

Spirituality and Deep Ecological Consciousness: Buddhist and Hindu philosophies on interrelationship.

"Deep Ecology" says Macy[12] (1994) is a term coined by Norwegian Philosopher Arne Naess to contrast with "shallow environmentalism", a "Band-Aid" approach applying piecemeal technological fixes for short term goals. Deep Ecology teaches us that we humans are neither the rulers nor the centre of universe, but are embedded in a vast living matrix and subject to its laws of reciprocity. Deep Ecology represents basic shift in ways of seeing and valuing, a shift beyond anthropocentricism.

Macy further considers the Tibetan and Sikkimese 'wheel of life' that mythically depicts all the realms of being. At the centre of that wheel of suffering are three figures: the pig, the rooster, and the snake- they represent delusion, greed and hatred- and they just chase one another around and around. The linchpin of all the pain is the notion of ourself, the notion that we have to protect that self or conquer on its behalf- or do something with it. This point of Buddhism, Macy thinks, is Deep Ecology too, in that we do not need to be doomed to the perpetual rat race. The vicious circle can be broken through wisdom, meditation and morality as practiced within realms of Mahayana Buddhism.

Hindu and Mahayana Buddhist philosophies followed within the study area, offer much to ponder on the spiritual and metaphysical connections between human and nature. The Hindu religion teaches that earth is the Universal Mother with the rivers and earth being personified as goddesses and the winds and fire as male deities (Chapple[13], 1993). Nature is the spiritual guide and teacher in a religion, which has codified environmentally beneficial practices in its rituals, proverbs and stories (Gadgil and Guha[14], 1992). For example respect and even reverence for other species is based on Hindu beliefs that the Supreme Being is incarnated in various species, and that human themselves can be reincarnated as animals and birds. In Hindu religion trees and plants are seen as possessing divine powers, and thus are to be protected from destruction.

Buddhism proposes that '*nirvana*' is a higher ultimate spiritual reality and truth which transcends all else. The 'Middle Way' of Buddhism teaches the importance of simplicity in living, that attachment and craving for material wealth and goods, stands in the way of liberation (Badiner[15] 1990). The Buddhist principle of the truth of personal experience emphasizes the importance of experiential learning, practicing the 'art of mindfulness' (described further in section on *chorten*s and monasteries to resensitize oneself to abstract ideas of reality). The Buddhist principal of reciprocity and interdependence is based on the belief that all events and beings are interdependent and interrelated in a universe which is 'a mutually causal web of relationship' (Kaza[16] 1993).

Buddhist relational ethics teach that moral actions are those, which are informed, sensitive to and respectful of all beings (sentient or not) within interdependent relationships. In both, Buddhism and Hinduism, violence to other beings within the biotic community is immoral, and given the percept of interdependence and self-destructive. The principal of non-violence is also central to both traditions' concept of '*Dharma*' which has been interpreted to mean 'moral duty', 'right action' and 'the truth about right living' (Bhatt[17] 1989). The Buddhist and Hindu doctrine of '*Karma*' (Fate or destiny) also teaches the importance of taking responsibility for moral action in all of one's relationship.

Religious monastic orders have also served as models for radical environmentalist proposals for ecologically sustainable communities. In a spiritual extension of 'Deep Ecology' concept, Devall[18] (1990) proposes that Buddhists develop ecocentric '*Sangha*' communities within bioregions. The wildlife in such regions is not only protected, but also fed and groomed towards being a climax biotic community-both botanic and zoological life.

The Case of Upper Rangit Basin: Sikkim

As enumerated earlier, Hinduism and Buddhism, both interlaced with animist spiritualism are the main religious traditions in the Sikkim's Upper Rangit Basin. Since the impact of popular urban culture has not been fully felt in the region, the old traditions and customs continue to dominate the life cycles of the inhabitants. There has been however a diffusion of cultural and religious ritual from within the inhabitants.

The Bhutias who immigrated from Tibet, follow the Lamaist Mahayana Buddhism diffused with the Bon animist spiritualistic traits. The Lepchas embraced Buddhism on coming in contact with the Bhutias, but retained their animist '*bon-thin*' and '*mun*' traditions. The Kiratas (Limbu, Rai, Gurung) embraced Hinduism through continuous interrelationship with the other Hindu Nepali castes. The traditional spiritual elements of Buddhism and animism can still be noticed amongst the Kirata communities. The other Nepali communities are mainly Hindus ; Tamangs and Sherpas being Buddhists. The Tamang community generally follows the Mahayana Buddhist religion, but their cultural traditions are of the kind of Nepali Hindu style. Both the religious groups have borrowed from each other to such an extent, that sometimes it is not easy to distinguish one from the other.

Sikkim was under the rule of Chogyals till the year 1975. The state religion during the Chogyal period was Mahayana Lamaist Buddhist . Although the followers of Hindu faith were numerically larger, the culture had greater impact of Buddhism and monastic life. The state character was therefore mainly Buddhist, which is evident from the architecture, laws, education, and the spiritual power centres - the monasteries.

The Lama's word has been taken as the law by the laymen. Almost every action and routine, including the rites during birth marriage, death, agricultural harvest, and festival celebrations are carried out by the Lamas.

In the sections of this chapter, the spiritual aspects of the *gompas*, monasteries and *chortens* will be discussed with amplified emphasis on the aspects of human ecology, spiritual ecology and deep ecology.

SECTION - II
ARCHITECTURE AND SPIRITUAL ECOLOGY OF MONASTERIES

Introduction

According to Risley[19] (1891) monasteries in Sikkim are of three kinds, namely:-

(a) *Tak-phu* (*brag-phug*) or cave -hermitage.

(b) *Gompa*, literally 'a solitary place', or monastery proper.

(c) The smaller '*gompas*' without the educational infrastructure found in larger Monasteries. The '*gompas*' as a rule, are merely temples (*hlakhang*) with one or more *lamas* engaged in ministering to the religious wants of the villagers.

The Upper Rangit Basin is unique for the fact, that it contains monasteries from the three major sects of Lamaist Buddhism viz *Nyingma, karma Kagyu* and *Gelug*. The *Gelug* monastery of the reformed school (of which the Dalai Lama is the functional head) is located on the Eastern end of the basin at Ravangla in the Kunphenling Settlement of Tibetan refugee camp. Ralang, the oldest monastery of Karma Kagyu sect in Sikkim is located at a 13 km distance from Ravangla, in full view of the *Nyingma* sect monasteries of Pemayangtse and Tashiding. Dubdi, the oldest monastery of Sikkim is located on top of Eastern banks of Rathongchu near Yuksom. A monastery belonging to the Bon practitioners (the religion of Tibet during pre Buddhist period) has also been constructed at a distance of 6 km from Ravangla on its Western slopes, on road to Kewzing- Tashiding. Hindu temples also exist in all villages and towns.

The Basin finds itself as a spiritual centre of Buddhism, Bon and Hinduism. It is probably the smallest geographical area housing monasteries from all sects except Sakyapa making the place spiritually unique.

Architecture of the Monasteries

The Buddhist '*gompas*' and *monasteries* in the Basin are characteristic of building tradition which came from Tibet together with the religion. However, at least three factors (GRIT library Gangtok Wall display)$_{20}$ have given rise to alteration that determine a style typical of Sikkim:-

(a) Sikkim especially the Rangit Basin was up to recent times very sparsely populated and relatively poor. Large buildings were not affordable and the building enterprises had to be adapted to the economy of the state. The monasteries were therefore initially more modest in size and decorations.

(b) Secondly the differences in altitude and wet climatic conditions of Sikkim necessitated a change in the types of roofs. The flat roofs of Tibet were replaced by slant ones, originally thatched from split bamboo, and later from corrugated iron galvanized sheets .

(c) Lastly, the sect of Buddhism that gained the firmest foothold was Nyingma- that is the unreformed school within Tibetan Buddhism and these gave rise to a type of monasteries that was radically different to that of Tibet, where the prevailing was of the Gelugpa sect.

Site Selection Methodology for monasteries

The monasteries in the Basin usually consist of one or several '*gompas*' (hlakhangs) depending on their size. They are surrounded by other smaller houses for accommodation of the monks during short stays. Meditation halls, kitchen and store rooms are constructed separately. The Ralang monastery has a large central courtyard and large dining hall.

The building site for such monasteries is selected with great care, with due considerations to auguries and omens. The courses of streams, the shape of mountains, the ground configuration and the colour of soil- all come into consideration and deep study prior to erection of the structures. All preliminary investigations are carried out not only for reasons of rituals, but also take into account the centuries old experience of generations, so as to avoid areas prone to earthquakes, landslides, floods, avalanches and wind hazards. Simple geo-technical tests are also carried out, such as digging holes in the soil and refilling them. This test reveals the density of soil and ensures that monasteries are not constructed over ground containing hollow space beneath the top layers. Likewise, if a hole in ground is filled with water, which sinks slowly, it is taken as a positively good sign.

Construction and Design

Wall Designing. Outer walls of the '*gompas*' are constructed from rough hewn stones laid in courses of alternate thick and thin sizes. The thick ones are called 'males' and the thin ones 'females' of human species. Cracks between the stones are filled with clay-clay mortar. In some '*gompas*' the entire wall surface way be plastered. The walls may be as much as one metre thick at the base, tapering as one reaches the top, the inner side being vertical all the while. All inner sides of the walls are plastered with clay.

Doors and Windows. The doors, windows, balconies and other similar wooden structures have a number of constructural details protruding from the façade. These are emphasized by painting relief carvings, giving the wood work an elevating effect, almost jolly, in contrast to the substantially slanting walls, which reflect an attitude of tranquility (SRIT staff interview, 2000).

Wall Painting. Colour of the masonery is pale greyish yellow, which may sometimes be white washed. All '*gompas*' have a broad horizontal band painted brick-red around the top of the walls , bounded on both sides by a plank painted with a carved motif of white pearls.

Roofs. As already mentioned, the roofs are hipped with small triangular gables placed vertically just under the ridge. The overhang may project up two or three metres, protecting the wall, and at the same time covering the prayer wheels, that usually circumference the temple at the base. As new roofing materials were introduced the shape of the roofs (as in the case of remodelled Pemayangtse and Tashiding monasteries) has gradually changed. The chubby profile of the thatches became more rigid when it was replaced by corrugated iron, and it was possible to make concave upward-swinging roofs as witnesed in Burma, Japan and China with flanged galvanized sheet iron. Mamy of the old '*gompa*' buildings were, in a way disfigured, taking away with it , the rusty but soothing ethnic look.

The Directional Aspects. The entrance to the '*gompa*'(*hlakhang*) faces East, and the altar is built up against the West wall. The '*gompas*' entrances thus receive the first rays of the rising sun. Access to the loft is from an outdoor stair case on the North side of the building, or via a ladder indoors.

The Interior. The larger '*gompas*' contain sizeable prayer halls down stairs with four or six column, and usually an entrance hall to the East. Some rooms are constructed in Northern and Southern directions for larger sized prayer wheel drums or for storage and a stair case leading to first floor. At the first floor again , the central hall has an altar at the Western end, and is generally arranged systematically. One of the side rooms on first floor is sometimes used for spiritual and *tantric* rites. The doors to the 'tantric rites' are painted with demon heads.

First and Upper Floors. The wooden structures supporting the first or upper floors are rather complicated. The corbelled pillars support a beam, upon which rest, first smaller, then larger cross beams. Since new materials have replaced the older ones, beams and floors are now cast in concrete. The top floor at Pemayangtse continues to support woodwork however.

Gompa features in Other Buildings. The older government buildings and some living houses of the '*Kazis*' adopted many features from the monasteries. The embellishment on the doors and windows was often the same, and also the construction of the beams and pillars. This is more common with the Bhutia houses and houses of some richer Lepchas. Economically less fortunate Lepchas however retained their original building traditions, usually simple huts made mainly from bamboo.

The Artisans

There are some artisans who are trained in the various skills of building, but they must often do other works besides this to support the family, usually farming. The more delicate decoration work in the temples is performed by specially trained monks.

Certain Designing Aspects

The sun and the moon designs mark the centre over the doors, windows or portals. These prevent evil from penetrating into the building. The row of white circles represents pearls as shown in Figure 3.1.

The traditional carving and decoration of the cantilevered capitals, symbolize the seven items necessary for the lord of the universe. Figure 3.2 depicts the design.

Prayerwheels

The prayerwheeels and drums contain strips of paper or cloth with prayers written on them, all of which are said to be sent up with each rotation. One does not pray for one's own sake, but prays for all other beings as well.

Fig. 3.1 Designs over Doors and Windows

The Sun and moon design marks the centre over a door; window or a portal. It prevents evil from penetrating into the building. The row of white circles represents pearls.

Fig. 3.2 Traditional Carving over Cantilever Capitals

The traditional carving and decoration of the cantilevered capitals, Symbolize the seven items necessary for the lord of the universe.

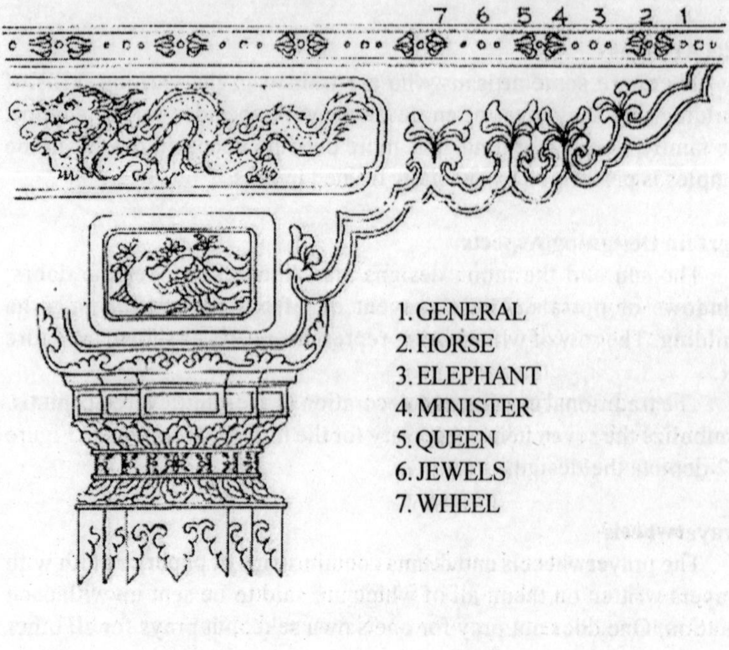

1. GENERAL
2. HORSE
3. ELEPHANT
4. MINISTER
5. QUEEN
6. JEWELS
7. WHEEL

Prayerwheels may function through the force of water, and are placed at man-made or natural water falls, where foot tracks or roads cross a rivulet or a fast flowing stream. Their blessings reach out to the travellers. Some prayer wheels function on wind energy as well. One such set of prayerwheels, functioning on wind energy is placed at the entrance of the Tashiding monastery.

Prayerwheels inside the '*gompas*' and some houses work on the principle of heat waves provided through lighted lamps or electric bulbs. A low pressure is built inside the wheel covering, which rotates due to gush of air from outside.

Four kinds of prayer wheels are shown in Figure 3.3

Prayer Flag

The prayer flags are erected outside the monasteries during various religious ceremonies. These flags are also erected outside houses and at the outer boundaries of villages, towns, localities and even the government buildings. The flags are differentiated on the basis of their colour pattern. The colours represent various elements found in nature, which are the following types:-

(a) Blue - Sky
(b) White - Iron
(c) Red - Fire
(d) Yellow - Earth
(e) Green - Wood (tree or Forest)

Aim of hoisting prayer flags is to bring good luck, peace harmony and to spread the holy messages. It is believed that the holy messages inscribed on these flags spread to distant places with the flow of wind thereby purifying the whole environment.

A single prayer flag can consist of all the colours or different flags of single colour can be erected. These colours are arranged in pattern with blue being on top followed by white, red, yellow and green at the bottom. Same pattern is followed from left to right if uni-coloured flags are erected.

Certain mythological figures are printed on the flags, which may be of following types (Also see Figure3.4) :-

(a) Divine horse (*Lungta*) symbolizing good luck.
(b) Buddha (*Gyantse-tsenmo*) design, mainly for bringing good luck to warriors. It also epitomizes law of the land.
(c) Guru Rinpoche (*Benza -Guru*)for curing ailments and for speedy recovery of health.

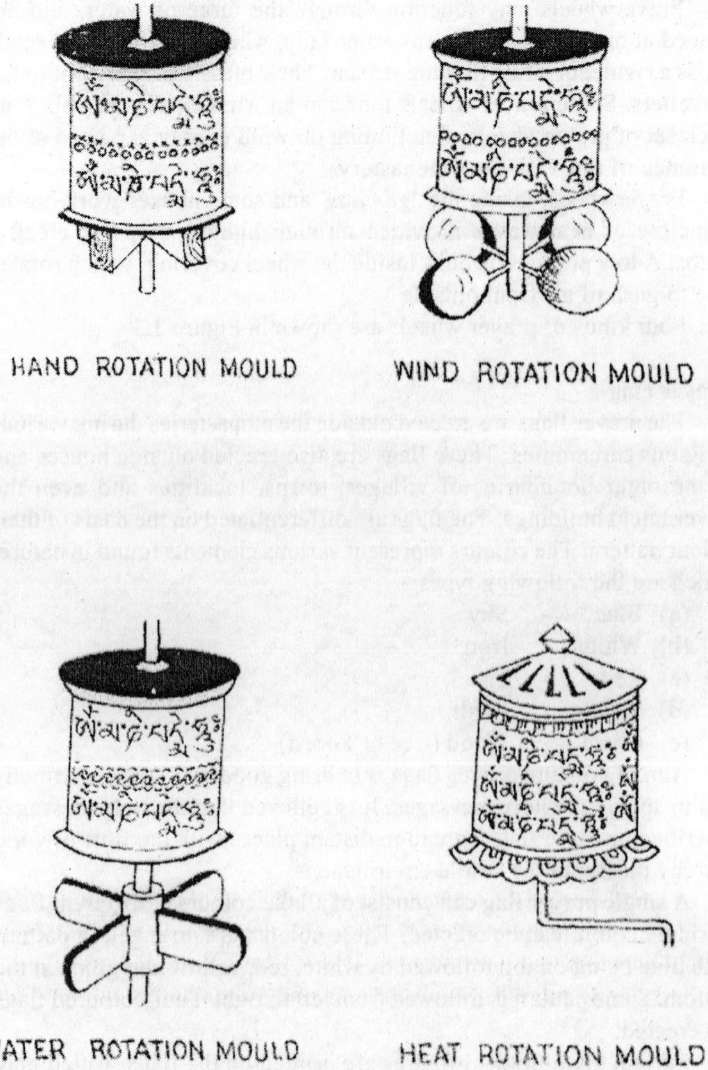

Fig. 3.3 Prayer Wheel

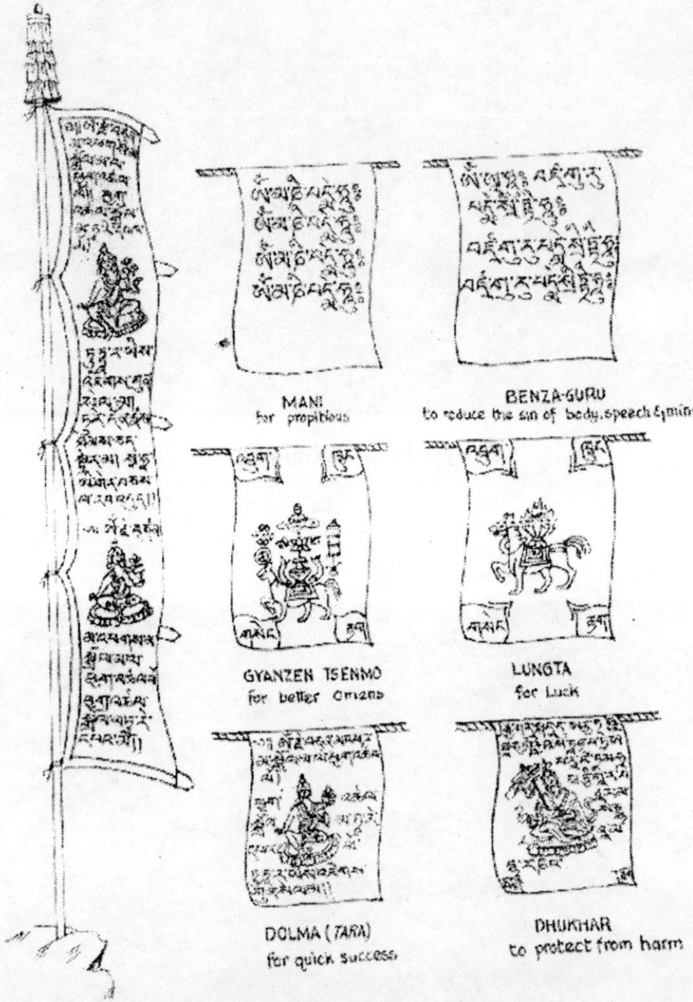

Fig. 3.4 Prayer Flags : Symbolic Designs
(*Mantras included here*)

Prayer Flags : Tashiding

Wind Rotation Prayer Wheels : Tashiding

(d) *Mani*, which is a form of Lord Vishnu for washing away any sin, which might have been committed while performing duties.
(e) *Dolma*, which is form of Goddess *Tara* symbolizing good luck.
(f) *Dorjee-Choepa*, which is one of the forms of Lord Buddha.

Type of flag to be hoisted is decided by the *Lama*, taking into consideration the occasion for which the flag is being erected, and also in some cases the horoscope of the individuals.

'Archaeology-Symbolism-Nature' Interrelationship

Nature and animal world form an integral part of the '*gompas*', including the symbols of prayers wheels and flags :-

(a) The monastery gates are decorated with deer paintings at the top. The wheel of life compares human spirituality and actions through the actions of animal life.
(b) Monastery construction itself commences after all aspects of natural order, and building sustainability are taken into account.
(c) The natures power of wind and water energy find their uses in rotation of the prayer wheels.
(d) The prayer flags depict all elements of nature through the colour combinations.

As such, natural forces are given full reverence, and the respect due to them in all prayer rituals. It is for followers to understand, and apply these lessons into their daily routine while interacting with the environment and the biotic life around them.

SECTION-III
TASHIDING MONASTERY: ITS HISTORY, AND STRUCTURAL LAYOUT-A CASE STUDY

History and Location

Tashiding monastery is considered to be the holiest of all the monasteries in Sikkim. It stands on a helmet shaped feature atop the confluence of Rathong chu and Rangit chu. Tashiding can be reached by road from Legship along a 17 km gravel road after crossing the majestic hanging bridge two km upstream from the Rangit hydel project. Legship itself is a trijunction of roads connecting it to Jorethang, Gyalshing and Ravangla. Though at a comparatively low altitude of about 4000 ft a.s.l, the feature stands like a majestic barge above the Rathong-Rangit confluence, which itself is marked by virtue of a serene *chorten* and a cave in vicinity. The location itself is spiritual. There is the Dubdi monastery 25 kms to the North, the wishfulfilling Khecheopalri lake to the North-West, the Pamayangtse monastery to the West, the Legship Shiva temple to the South, the Mongbrue *gompa* and Ravangla Bon monastery to the South-East, the Ravangla Gelug monastery to the East and the Karma Kagyud Ralang monastery to the North-East, making Tashiding virtually the spiritual centre atop the base, wherefrom geographically the Upper Rangit Basin takes off towards the Kanchandzonga massif and the two water sheds of Singalila and Mainam-Tendong ranges. Barring Dubdi and Kheecheopalri, all other features named above are visible from Tashiding.

Layout of Tashiding Market, Monastery and *Chortens*.

During the course of field work in addition to historical and spiritual research, a ground relief survey was also carried out. Commencing from the Tashiding (Sinek) market place and moving southwards, the ground configuration was divided into five parts. All five parts have been depicted on five different plates in Figure 3.5 as under :-

(a) **Plate I** Shows the Sinek market place. It is essentially a linear settlement, almost self sufficient in terms of daily needs. The settlement is located on a dip over the ridge between Rathong chu and Rangit chu. On its North at a distance of 3 km stands the Sinolochu '*gompa*', and the Southern extension leads to Tashiding monastery. Yoksom is 23 km to the North-North west.

(b) **Plate II** depicts the outskirts of Tashiding (Sinek) market on its South. The large '*Mani*' inscribed on a huge boulder dominates the entrance to the settlement.

(c) **Plate III** is the depiction of the area from the market place upto the monastery entrance gate. One can approach the monastery by a light vehicle on a narrow gravel road as shown on the plate, or move on the foot track, which is on a gradual slope rising Southwards. Near the '*Mani*' at track junction, one foot track branches off, which by outskirting the Tashiding feature leads to a cave in the South. Just short of the entrance, on the right hand side of the foot track the 'prayer wind wheels' whisper their prayers into the ears of the perspiring visitors, rejuvenating them physically as well as spiritually.

(d) **Plate IV** depicts the main monastery area, a part description of which has been given at the beginning of this section. The main entrance is marked by 'Mani Lhakang', followed by a cluster of prayer flags. Further to the right, the guest rooms are laid out. These are followed up successively, by the monumental '*Tashiding Gompa*', the 'butter lamp' house, four *chortens*, Tsenkhang, a new 'butter lamp' house, and finally 'Guru Lhahang' the temple of Guru Rinpoche. Kitchen, school and private quarters lie on the left side of the track. The constructional details and corresponds to the details given under the preceding section dealing with ' Architecture and Spiritual Ecology'.

(e) **Plate V** shows the *Chorten* area, which are 41 in number. Three kinds of *Chortens*, i.e enlightenment, reconciliation and great miracle are noticed. A detailed note on the *Chortens* is given in another section of this Chapter. The 'rinpoches' and 'tathagatas' to whom some of the '*Chortens*' correspond to, is given at page 112.

Spiritual Beliefs

A number of legends are associated with the site, where the present monastery stands. Guru Rinpoche 'Padmasambhava' is believed to have stayed here for meditation. Legend has it that he shot an arrow, vowing to meditate at the place where it fell. It dropped on the place where the present monastery is located. It is also believed that the site at Tashiding was accepted as the most sacred in Sikkim by the first three monks following the observance of occurrence of the divine phenomenon. A bright streak of light issuing from the top of Mount Khangchendzonga is believed to have shone right upon a spot at Tashiding. Simultaneously a most delicious fragrance of incense and notes of melodious music filled the atmosphere. The monks accompanied by King Phuntshog Namgyal visited the spot, and built the *chorten* named 'Thongwa-Rang-Grol', the sight of which confers self-emancipation. (Namgyal and Dolma[21], 1908).

The Ngadak Lama built the monastery of 'Tashi-ga-legs' and also a Maitreya Temple at the site. The Chogyal Lhakhang was built by Pema Wangmo, elder sister of king Chagdor Namgyal during first half of 18th century. In 1715 Jigme Pao and King Chagdor Nangyal constructed Guru Lhakang, and also supervised the internal furnishing. Jigme Pao himself performed the consecration ceremony of all Lhakhangs and *Chortens* at Tashiding. (Dhamala[22], 1993).

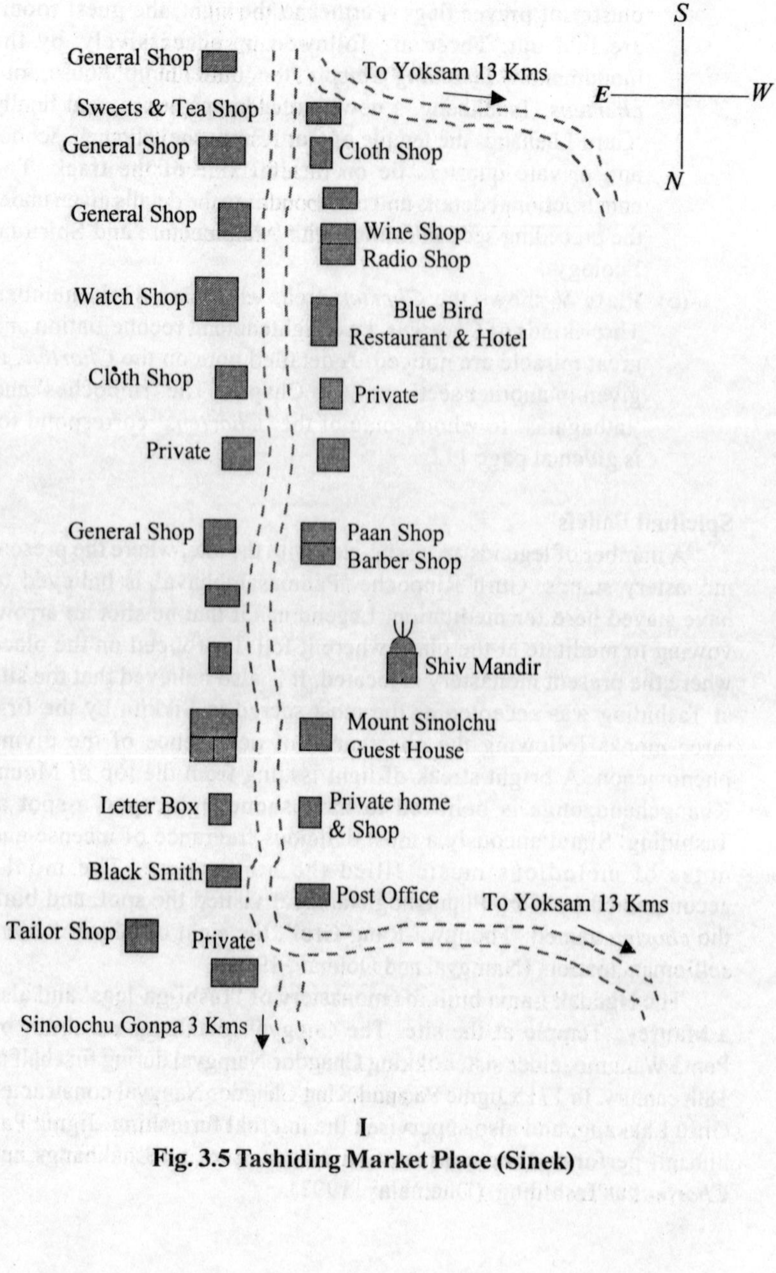

Fig. 3.5 Tashiding Market Place (Sinek)

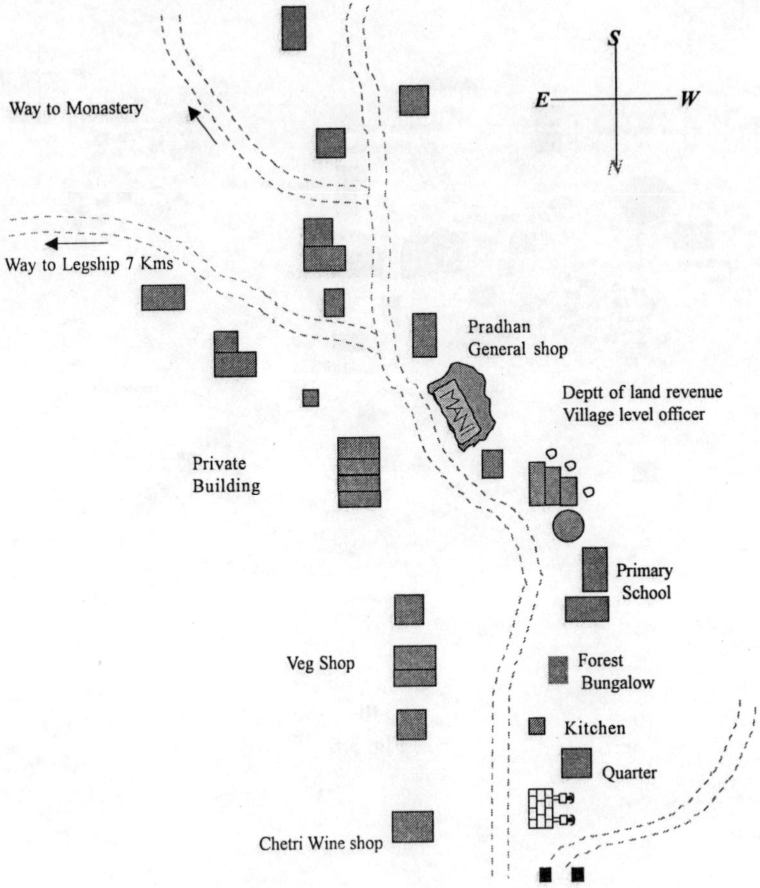

II

Fig. 3.5 Outskirts of Tashiding Market

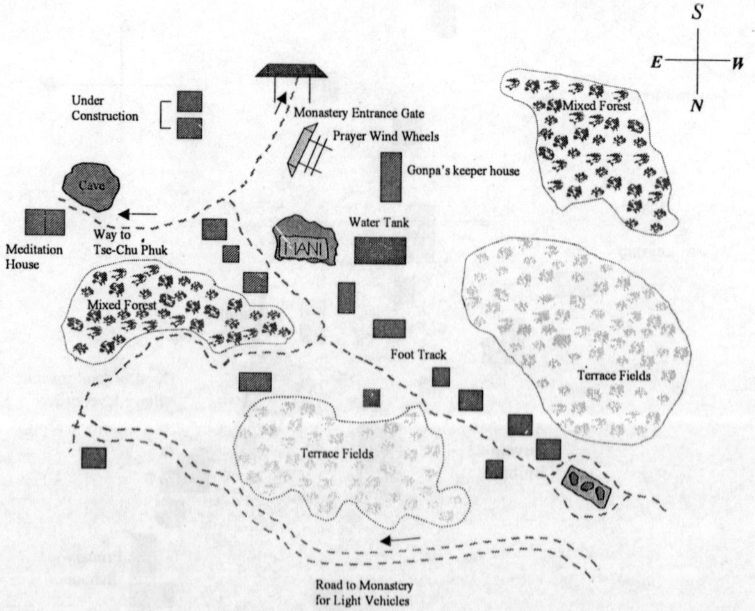
S
E
W
N
Under
Construction
Monastery Entrance Gate
Prayer Wind Wheels
Gonpa's keeper house
Mixed Forest
Cave
Way to
Tse-Chu Phuk
Meditation
House
Water Tank
MANI
Mixed Forest
Foot Track
Terrace Fields
Terrace Fields
Road to Monastery
for Light Vehicles

III
Fig. 3.5

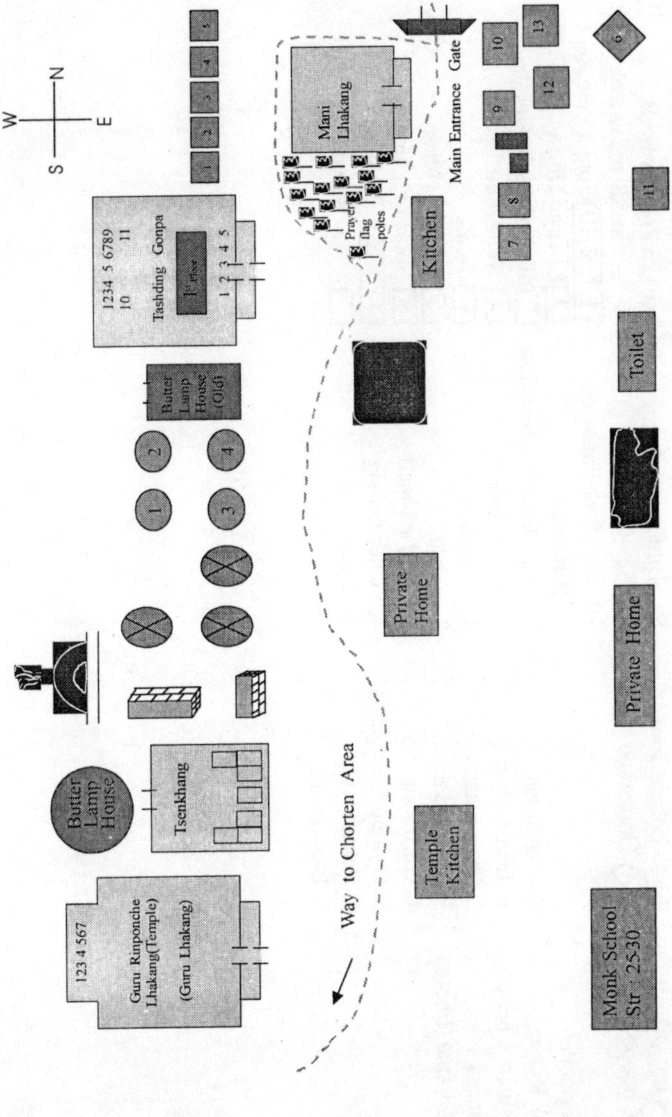

IV

Fig. 3.5 Monastery Area

Note :-

Guru Lhakhng
1. First King Chogyal Phuntsok Namgyal
2. Lhapte Chenpo
3. Buddha
4. Guru Rinpoche
5. Phuntsok Rinzin
6. Kadok Rinzin Chenpo
7. His Holiness The Dalai Lama (Photo)

Chortens
1&2. Unknown
3&4. Enlightenment
⊗ Unmarked

Tsen-Khang
1. Dham-chen
2. Tsen-maar
3. Tsen-maar
4. Guru Rinpche
5. Gyalpo Ku-nga
6. Pehar Gyalpo
7. Dam-chen

Tashiding Gonpa
1-4 Ser-Chengyed
5. Buddha Statue
6-9 Ser-Chengyed
10. Ngadhak Phuntsok Rinzing
11. Ngadhak Rinchen Gon (First floor)
1. Rinchen
2. Guru
3. Hod-pa-med
4. Chenren-sig
5. Guru

Manilhakhang
(i) Chenren-sig Statue
(ii) 23 Prayer wheels inside the wheels

[1,7,9&10] Gonpa's House
[2] Khenpo House
[3] Six Rooms for Guests
[4] Nun Guest House
[5] Private
[6] Teacher House
[8] Sponsor's House
[11,12&13] Private
[ ] Water Tank

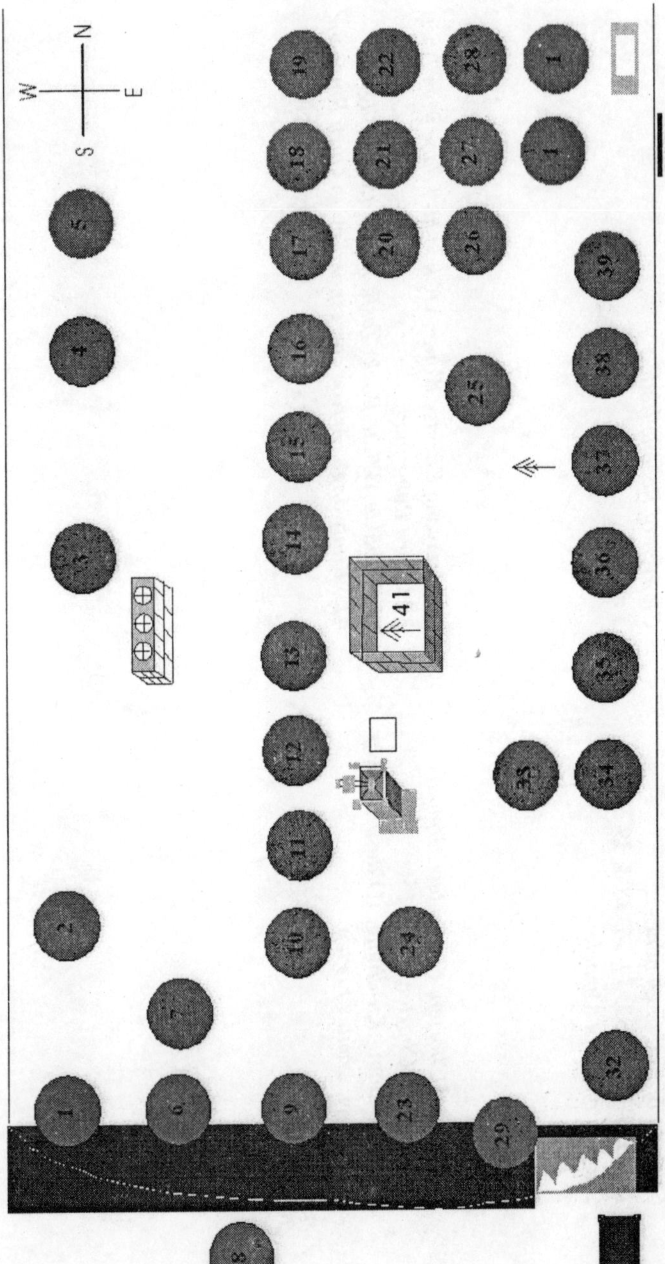

Fig. 3.5 Main Chorten Area of Tashiding Monastery

Types of Chortens

Chorten of Enlightenment:-
1-4, 6, 8-17, 19-32, 34-36, 38 & 39
Chorten of Great Miracle:-
7, 18 & 33

Chorten of Reconciliation :-
5 & 20

Note:-

8 Dhudal Chorten. (To protect Gonpa)
24 Lhabab Chorten
25 Woe-suung Ku-dhung. (Oldest/Largest)
31 Kagyu Lama Chorten.

32 Jamyang Khentse Wongpo (Choe-ki- Lodoe from Tibet)
37 Guru Rang-jong
40 Dhakar (White Rock) Tashi Dhing (Main pilgrim)
41 Khatwanga, Arrow corresponding to Guru Rinpoche

Legend and Anecdotal Accounts

A conversation with the monks at Tashiding, and a study of literature at Sikkim Research Institute of Tibetology, Gangtok revealed the history, legends, and anecdotal accounts attached with the spirituality of Tashiding monastery. True or not, the messages emanating out of the accounts reveal the Sikkimese people high esteem and faith in spirituality, leading to *Karma*, which is central to balance of nature or 'Deep Ecology'.

According to the Neyig (Guide Book to the Holy Places) Beyul Demojong (Hidden Valley of Rice) had five great provinces and six hidden spots which protected all living beings. Guru Rinpoche concealed innumerable scriptures (*Vhos*), Wealth (*Nor*) and sacred objects (*Wangten*) in those holy places and performed many inexpressible benedictory prayers for the benefit of sentient beings thereby entrusting all the treasures to the protectors and tutelary deities for their preservation. He blessed and consecrated this land as sacred as Urgyen Zangdogpalri and Kanchopalgiriwo, the realms of unlimited happiness and abode and assemblage of female divinities (*Mamo Khandos*) like clouds in space.

Likewise the innumerable Buddhas and Bhodisattavas of ten directions also blessed the land. It is believed that one who visits these sacred lands, the doors of hell would be closed for him i.e. they become immortals. Tashiding which is one of the most sacred places of Sikkim has four holy caves in four different directions. They are Sharchog Bephug in the East, Khandozangphu in the South, Dechenphug in the West and Lhari Nyingphug in the North. The centre of the holy place is known as Dakkar Tashiding, which is also the main deity of Tashiding.

During the first dissemination of Buddhism in Tibet in the 8th century A.D the illustrious Guru Padmasambhava and twenty-five disciples (*Je-bang nyer-nga*) had blessed the land (*Beyul Demojong*). Thereafter they tamed all malevolent spirits and evil forces which infested this country. They also erected a number of places of worship, such as monasteries and *chorten*s and the land was thus transformed into a great sacred and holy place.

In course of this process of propagation of Buddhism into Sikkim two great Siddhas (Tertons-revealers of Tantra) Rinzin Godemchen (1340-1396) and Kathog U, o'd Yeshe Bumpa came to Sikkim and established monasteries and meditation centres at sacred places of 'Pawo Hungri' and Tashiding. The reincarnation of Rinzin Godemchen, Terton Ngari Rinzin Chenpo Lagdendorjee discovered the sublime tantra called Khadoi Khalangma from the cave of Lhari Nyingpo.

The seventeenth century was the turning point in the history of Sikkim as the first Chogyal Phuntshog Namgyal was enthroned at Yuksom by three great Lamas of Tibet namely Lhatsum Namkha Jigme (the embodiment of compassion was known to be the incarnation of Indian Maha Pandita Vimalamitra and Tibetan omniscient Kunkhen Longchenpa Dema Hozer), Ngadag Sempachenpo Phuntshog Rigzin and Kathog Rigzin Chenpo. Under the Patronage of King Phuntshog Namgyal, monasteries, hermitages for meditation and *chorten*s were built. Chogyal Phuntshog Namgyal proclaimed Buddhism as the state religion.

Through the revelation of Zogchen system of meditation and Sadhana, Sikkim became the main centre of Zogchen teaching. Its followers came to be called, Denzong Zogchenpa.

The two great scholars Ngadag Sempachenpo and Ngadag Rinchen Gon established the holy shrine at Dakkar Tashiding and consecrated it to be the principal seat of the followers of Nyingma sect in Sikkim.

Holy Water Vase, and Bum Chu Origination

The belief as to how the celebration of 'Holy water Vase' festival originated is briefly narrated hereunder. During the reign of the religious king Trisong Deutsen in Tibet, Guru Padmasambhava , while bestowing the unparallel tantric system of tutelary deity Mahakarunika Avalokitesvara Sadhana and initiation on emancipation from the cycle of mundane existence to king Trisong Deutsen, prince Murub Tsenpo, Yeshe Tsogyal and Verotsana , consecrated the same Holy Vase which is preserved in the Tashiding monastery till today.

According to the works of Zhigpo Lingpa, the Holy Vase is made of various kinds of sacred soil, water and five kinds of precious jewels (Rinchen Na -nga) collected from the holy places of India, Odiuana and Zahar by Guru Padmansambhava. It is said that the Vase is made by Damchen Gar-bgag, wrathful Dharmpala and consecrated by Guru Padmasambahva by conducting Sadhana of Yidam Chuchig Zhal (tutelary deity of eleven heads). When he performed the sadhana, the entire retinue of the deity and deity himself are said to have appeared in the sky and inseparably merged into the Holy Water contained in the Vase. Thereupon the holy water overflowed and spread in all directions in the form of rays. That very moment, as a sign of good omen, there was earthquake, the four guardian deities (Gyalchen Dezhi/ Cutur -Maharajika) of Dharma and the gods of thirty-three heavens (Sumchu Tsasumgyi Lhanam) showered flowers from the sky. This event was witnessed by the people assembled there. All the people were overjoyed at this spectacular event and Holy Water was distributed to the devotees from the Vase. Yet it was found that the holy water of the vase never decreased. Inspired by this, the sinners became virtuous, the pious people realized the lofty divine qualities and all were spiritually benefited. Finally Guru Padmasambhava concealed the 'Holy Water Vase' into the sublime hidden treasure and entrusted it to the protective deities.

In the later period, the great Terton (revealer of hidden treasures), Chokyi Gyalpo Garwang Rigzin Zhinpolingpa, the reincarnation of prince Murub Tsenpo of Tibet, unearthed the Vase of Holy Water for the sake of sentient beings. After the discovery of the Vase, Terton Zhipolingpa maintained it as his Thugdam Ten and in the later part of his life, he offered it to Terton Tagshamchen with special instruction. Terton Tagshamchen then handed over the 'Holy Vase' to Ngadag Sempachenpo Phuntshog Rigzin to be kept and installed at Dakkar Tashiding, the holiest and most blessed hidden land of Guru Padmasambhava. So Terton Ngadag Sampachenpo brought the Vase to Tashiding and installed it in the Tshuglagkhang. He is known to have conducted special recitation of 1,300 million syllables 'OM MANI PADME HUM' through Thugjechenpo khorwalegrol Sadhna under the royal patronage of the first Chogyal Phuntshog Namgyal. At that time many unprecedented and auspicious signs had appeared in the country.

Bumchu celebration is one of the most important and holiest events in Sikkim and thousands of devotees from all over the state and its neighbouring countries come on pilgrimage and participate in the holy occasion. The Vase, containing holy water, was kept in a miniature mansion (cho-sham) under lock and seal of the Chogyal of Sikkim Every year special recitation is conducted and the seal is checked by the high officials and Lamas before the Vase is taken out of the mansion. The Vase is opened on the night of the 14th day of the first month of the Tibetan Almanac. From the Vase three cups of the water is first taken out. The first cup is meant for the royal family, the second cup for the Lamas and the third is distributed among the devotees on the full moon day. Fresh water is brought from Rathong Chu, which is also considered as a blessed river, as replacement.

In some years the Holy Water in the Vase increases by 21 cups and other times it decreases. In some cases the water of the Vase remains at the same level but other times it is found dusty. These are believed to be the predictions of auspicious and inauspicious occurrences in the country. Wehn the water level increases, it is a sign of prosperity and when it decreases it is an indication of bad year of drought and diseases. The dusty water indicated conflicts and unrest.

Summary

For Tashiding, one can say 'Seeing is believing'. The monastery is historically illustrious, geographically well located, aesthetically beautiful, spiritually divine-a place where nature and spirituality dwell together, urging the human race to be ecologically upright.

SECTION-IV
CHORTENS: ARCHITECTURE, SPIRITUAL AND ECOLOGICAL SIGNIFICANCE

Introduction

The mysteries of life and death were always the greatest agents of religious ritual and speculation. Through the experience of death man becomes conscious of life. Thus the cult of the dead stimulated primitive man to build the first great monuments (tumuli), while the other side of religious activity, which was concerned with the living and the mundane aspects of life, found expression in the simpler forms of tree and fire-worship. The tumuli originating from the burial mound, were massive structures of stone, taking the forms of hemispheres, cones, pyramids and similar plain stereometrical bodies, containing small cells which preserved the bodily remains and other relics of heroes, saints, kings and similar great personalities. In India, as in many other parts of Asia, the hemispheric form seems to have been the prevalent type of such monuments. According to the oldest tradition they were erected for great rulers and sages. (Govinda[23] 1970).

While the tumuli and the cult of the dead had their place outside the village, the sanctuary of the life-giving and life-preserving forces (personified in the sun-god) has its place in the centre of the village. It consisted of a simple altar)a sanctified form of the domestic hearth, the fire of which was always regarded sacred as a symbol of family life) or as small shrine (an idealized form of the village hut) which stood in the shadow of the sacred tree (the Tree of life) and was surrounded by a fence as a demarcation of the sacred place.

The Buddhist *stupa* combined the elements of the village sanctuary with the monumental dome of the ancient tumulus (*caitya*), thus uniting the two oldest traditions of humanity, as expressed in the lunar and solar cult, fusing them into one universal symbol which recognized formally for the first time that life and death are only two sides of the same reality, complementing and conditioning each other. To think of them as separate is illusion, and only as long as the veil of Maya has not been lifted, the worship of these two forces proceeds separately, sometimes even as two separate forms of religion. But once it has been understood that there is no life without transformation, and that the power of transformation is the essence of life-then the great synthesis takes place and the foundation of a world-religion is established. (Yeshe and Zopa Rinpoche[24] 1982).

Harmika

The Buddhist *stupa* originally consisted of an almost hemispherical tumulus and an altar-like structure (*harmika*) on its top, surmounted by one or several superimposed honorific umbrellas. The flattened hemisphere was compared to an egg and therefore called '*anda*', a terms which did not only allude to the shape (which was also compared to a water-bubble) but to its deeper significance as well, namely, as a symbol of latent creative power, while the quadrangular *harmika* on the summit of the cupola symbolized the sanctuary enthroned above the world (*anda* was also a synonym of the universe in the oldest India an mythology) beyond death and rebirth. A similar parallelism exists between the *harmika* in the shade of sacred tree, because the Holy One, whose ashes were enshrined in the altar like sanctuary of the *harmika*, instead of sacrificing other beings, had sacrificed himself for the welfare of all living beings. According to the Buddha there is only one sacrifice which is real value, the sacrifice of our own desires, our own 'self'. The ultimate form of such a sacrifice is that of a Bodhisattva, who renounces even the ultimate peace of final *nirvana* (*Parinirvana*) until he has helped his fellow-being to find the path of liberation.

The Monorific Umbrella

It is an abstract representation of the shade-giving tree in this case representing the sacred Tree of Life - is one of the chief solar symbols, and in Buddhism that of Enlightenment (*samyak-sambodhi*). The importance of this symbol becomes clear from the Buddhist Scriptures, describing the struggle of the Buddha and Mara, the Evil One, for the place under the Bodhi Tree, the holiest spot in the world, later known as the Diamond Throne (*vajrasana, Tib. rDo-rje gdan*).

It must have been an old custom that the head of the community had his seat of honour under the sacred tree on the centre of the settlement where public meetings used to take place on religious and other important occasions. Consequently the umbrella, which replaced the tree when the head of the community moved about or presided over similar functions in other places, later on became one of the insignia of royalty. In order to mark the distinctions in rank the ceremonial umbrella was doubled or trebled, or increased by even greater number of umbrellas, which were fixed one above the other, thus transforming the umbrella back again to the original tree-shape with its numerous layers of branches spreading around the stem and gradually getting shorter towards the top. These layered umbrellas are the thirteen rings or wheels along the entire length of the cone placed on top of '*harmika*' and '*pema*'.

Historical Backdrop

In order to understand the transformation of the ancient customary tumulus into the universal conception of the *Caitya*, from which later the Tibetan *Chorten* developed, we must have a look at the earliest known Buddhist Stupa at Sanchi. The great Sanchi Stupa was crowned by a threefold honorific umbrella and the altar-shrine on top of the hemispherical main structure was surrounded by a railing (*vedika*), exactly as in the case of the village sanctuary. Similar railings were repeated at the foot of the *stupa* and on the low circular-terrace upon which the flattened hemisphere rested. The lowest railing was provided with four gateways (*torana*) which opened towards the East, the South, the West & the North, emphasizing the universal character of Buddhism which is open to all the four quarters of the universe and invites all humankind with the call "Come and see", and which exhorts its followers to open their hearts to all that lives, while radiating love, compassion, sympathetic joy and equanimity towards the whole world. The inner space between the stone railing and the *stupa*, as well as the circular terrace (*medhi*) at the base of the cupola were used as *pradaksina patha* for ritual circumambulation in the direction of the sun's course. The orientation of the gates equally corresponds to the sun's course to sunrise, zenith, sunset and nadir. Just as the sun illuminates the physical world, so does the Buddha, the Enlightened One, illuminate the spiritual world. The Eastern gateway represents his birth, the Southern (which was regarded as the most important and therefore built first) his enlightenment, the Western his "setting in motion the wheel of the Law"(*dharma cakra pravartana*) : the proclamation of his doctrine, and the Northern his final liberation (*pariniravana*).

This universal attitude and orientation remained one of the characteristics of the *stupa*, especially in the Northern countries of Buddhism, like Tibet, even after railings and gateways had disappeared. In the course of time all these details were fused into a quadrangular substructure, which finally took the form of four terraces (sometimes furnished with four staircases, if the size of the monument permitted or required them) upon which the hemisphere was raised.

As the layers of superimposed umbrellas became more numerous they were transformed into the more architectural shape of a solid cone with a corresponding number of horizontal notches, which finally amounted to thirteen. With this transformation the original idea of the Tree of Life and Enlightenment was visibly restored and steadily gained in importance. That the conical spire was no more regarded as a set of umbrellas, can be seen from the fact that later on an honorific umbrella was again fixed on top of the cone. The succeeding section furnishes the details.

The different strata of the cone (separated by horizontal notches) were now explained to correspond to certain psychic faculties or stages of consciousness on the way to enlightenment and to their respective world-planes. Thus the spiritual rebirth of the world starts in the mind of man, and the Tree of Life grows out of his own heart, the centre of his world, and spreads into ever new infinities, into ever higher and purer realms, until it has turned into a Tree of Enlightenment.

Stupa and Individual Spirituality

Nepalese *stupas*, which in many respects have preserved archaic features, decorate the *harmika* (the cubic structure above the cupola) with painted human eyes, thus suggesting a human figure in the posture of meditation hidden in the *stupa*: the crossed legs in the base, the body up to the shoulders in the hemisphere, the head in the *harmika*. This also corresponds to the psycho-physical doctrine of the centres of psychic force (*cakra*) which are located one above the other in the human body, and through which consciousness develops in an ascending order: from the experience of material sense-objects through that of the immaterial worlds of pure mental objects, up to the supramundane consciousness of enlightenment, which has its base in the crown-*cakra* of the head (*sahasrara*). This *cakra* is symbolized by a dome-shaped or flame-like protuberance on the head of the Buddha, and by the cone-shaped Tree of Enlightenment which forms the spire of the *stupa* or the *Chorten*, or its various equivalents, like the *dogobas* (*dhatu-garbha*) of Sri Lanka or *pagodas* (a reversal of the word *dogoba*) in Burma, Thailand and Indo-China.

The *cakra* itself is a sun symbol. It was one of the attributes of the sun-god, either in the form of a discus or in the form of the wheel, representing the rolling sun-chariot. The solar origin is testified by the description of the flaming and radiating wheel which appears in the sky with its thousand spokes (rays) when a virtuous ruler has established a reign of righteousness and has attained the spiritual power which entitles and enables him to extend the Good Law (*dharma*) over the whole world and to become a world-ruler (*cakravartin*) . Similarly, the "turning of the Wheel of the Good Law" has become a synonym for the Buddha's first proclamation of his doctrine (*dharma-chakra-pravartana-sutra*), by which the thousand-spoked sun-wheel of the universal law was set in motion, radiating its light throughout the world. (Govinda 1960[25]).

Psychic Forces and Spirituality

The *cakras* as radiating centres of psychic force gave a new impetus to the interpretation of the human body as a cosmic manifestation. Not only was the spinal column compared to Mount Meru, the axis of the universe, and therefore called '*meru-danda*' but the whole psycho-physical organism was explained in terms of solar and lunar forces, which through fine channels, the so-called veins or nadis, moved up and down between the seven *cakras*, which in their turn represented the elementary qualities of which the universe is built and of which the material elements are only the visible reflexes.

Unity of Spirituality and Nature Forces : Chorten and Human Ecology

A parallelism was established with respect to the psychic organism whose vital centres (*cakras*) were found to correspond to the elementary qualities of matter; the basic vital centre or "root support : (*muladhara-cakra*), situated in the perineum at the base of the spinal column, (which latter represents the Tree of Life) and corresponding to the element Earth, the solid state; the navel-centre (*manipura-cakra*) to the element Water, the fluid state; the heart-centre (*anahata-cakra*) to the element Fire, the heating, incandescent or radiating state; the throat-centre (*visuddha-cakra*) to the element Air, the gaseous state;

and the centre on the crown of the head (*sahasrara-cakra*) to the element Ether (or in its passive aspect; Space), the state of vibration.

Each of these elements is symbolized by a sound (bija-mantra, a mystic syllable of creative power), a colour and a basic form. The latter two are of special interest to us, as they have been directly applied to the architecture of the *chorten* . Earth is represented by a yellow cube, Water by a white sphere or a white round pot, Fire by a triangular body of either round or square base; i.e, a cone or, less frequently, a pyramid. Air is represented two dimensionally as a semi-circular bow-shaped form of green colour, three-dimensionally as a hemisphere with the base upward, like a cup. Ether is graphically represented by a small-acuminated circle or blue dot (*bindu*) and appears in three-dimensional form as a multi-coloured flaming jewel, i.e, a small sphere.

If we put all these elements together in due order, namely, the sphere upon the cube, a cone or a pyramid upon the sphere, and upon the cone or pyramid a cup-like hemisphere which carries a flaming drop on its plane surface-then we get the ideal figure or the abstract stereometrical form which represents the basic principles of stupa-architecture, as preserved in the Tibetan *chorten* as well as in the Japanese *Sotoba* into a vase-or pot-shaped vessel (Tib : *bum-pa*) which rests on a cubic substructure and is crowned by a tall cone, ending in a small upturned hemisphere, which carries on its plane surface a crescent, a sundisc and the 'flaming jewel', one upon the other (Dorjee[26] 1996). This represents all elements of universe and human body together, sanctifying the sum total-an idea wholly supporting the concept of 'Deep Ecology', the biocentric spiritual aspect of human ecology. It is further amplified in succeeding paragraph.

The main parts of the *Chorten* are generally given the colours of the 'great elements' (*mahabhuta*): the cubical sub-structure yellow (Earth); the post-shaped central part white (water), the conical spire red (Fired), while the fourth element (Air) which should show a green surface, is generally hidden under the honorific umbrella, a symbol which, especially in its Tibetan form, is closely connected with a concept of Air. Without taking into account its tree-origin and its natural relationship to sun, air and sky. It may be mentioned that according to the later Indian and Tibetan tradition honorific umbrellas were supposed to appear in the sky, when a saint had realized certain magic powers. Between the umbrella and the flaming drop (Tib :*thig-le*), the respective symbols of Air and Ether, there is a white crescent, in whose inner curve rests a red sun-disc. They repeat the colours of the two main elements of the *Chorten*, namely that of the moon-related, waterpot-shaped central part and that of the sun-related conical spire. The meaning of this repetition becomes evident if we remember the role of the lunar and solar forces moving through the main channels or *nadis* of the psycho-physical organism of man. The most important one runs through the spinal column and is called *susumna* (*dhuma-rtsa* in Tibetan), while *ida* (Tib.; *rkyang-ma rtsa* and *pingala* (Tib.: *ro-ma-rtsa*) coil round the central channel in opposite directions, the pale white-coloured *ida* starting from the left (or, according to Tibetan tradition, controlling the left side of the human body), the red coloured *pingala* from the right (or controlling the right side). *Ida* is the conductor of the lunar or moon-like'(*candrasvarupa*) forces, which have the regenerative properties and the unity of undifferentiated subconscious life, as represented by the latent creativeness of seed, egg and semen, in which all chronic-telluric cults are centred. *Pingala* is the vehicle of solar forces (*surya-svarupa*),which have the properties of intellectual activity, representing the conscious, differentiated individualized life. Individualisation, however, if separating itself, from its origin, is as death-spelling as knowledge severed from the sources of life. This is why *wisdom and compassion (prajna and karuna) must be united for the attainment of liberation.* And for the same reason *pingala* the solar energy, without the regenerating influence of *ida*, the lunar energy, acts like a poison, while even the elixir of immortality (*amrita*), to which the regenerating lunar energy is compared, has no value without the light of knowledge.

It is for this reason that only when the solar and lunar energies are united in the central channel, the *susumma*, and carried up from the root-centre (*muladhara-cakra*) through all the other centres of psychic power and consciousness until they reach the universal level in the Thousand-Petalled Lotus of the *sahasrar-cakra*. The final integration of these two forces takes place and results in the ultimate state of illumination (*samyak-sambodhi*). In the spherical and conical part if the *Chorten* the two currents of psychic energy are represented by their separate and elementary aspẹcts; in the crescent and the sun-disc they are represented in their sublimated or spiritualized form as knowledge (*prajna*) and compassion (*karuna*), from the union of which the dazzling flame-jewel of perfect enlightenment is born. This symbol of unity and ultimate reality has its latent counterpart in the form of a blue dot (*bindu; Tib.: thig-le*) or seed (*bija*), the creative germ or spiritual potentiality, inherent in every sentient being as the potential consciousness of enlightenment (*budhicitta*; Tib.: *hyahng chub-sems*). The unfolding of this latent principle is the aim of the spiritual path, which is achieved when all our psychic faculties as embodied in the various centres- are permeated by it. When the mystic union between the sun of knowledge and the moon of compassion has reached its zenith and consummation on the highest spiritual plane, the Thousand -Petalled Lotus, then it comes to pass that the dark seed, containing the essence of the universe and the ever-present reality of the *dharma-dhatu*, breaks open and bursts forth into the dazzling flame of enlightenment, the crowning symbol of the most universal type of the Tibetan Mahayana Chortern (Govinda[27] 1970).

Conclusion

Combination of universal elements and forces, and uniting them with life system psychic powers and than placing them in the form of a *Chorten* urges the living humans to understand and have '*karuna*' for all existence-biotic or abiotic. This is encompassed the following quotation from Chandogya Upanisad :-

'The essence of all beings is earth,
the essence of earth is water,
the essence of water are the plants,
the essence of the plants is man,
the essence of man is speech,
the essence of speech is Rgveda,
the essence of the Rgveda is the Samaveda,
the essence of the Samaveda is the Udgita (which is OM)
That Udgita is the best of all essences, the highest,
Deserving the highest place, the eighth.

(CHANDOGYA UPANISAD)

It is essentially 'knowledge of' and 'compassion for 'all universal existence that leads to enlightenment. This message of '*Karuna*' and '*gyan*' is reflected in the '*chortens*.' This is human ecology in its deepest realms. This is the message of Buddha in the '*chortens*'.

SECTION-V
CHORTENS OF TASHIDING

Introduction

The *chortens* seen in all parts of Sikkim follow Tibeto-Buddhist ritual and architecture. Mention of various types of *chorten* architecture is incomplete without making references to the eight *stupas* associated with the eight major events in the life of Lord Buddha as under:-

(a) The first stupa was constructed in the Lumbini garden of Kapilavastu where the Buddha was born.
(b) The second stupa was on the banks of the Naranjana River in the kingdom of Magadha, where the Buddha had attained enlightenment.
(c) The third stupa was constructed in the Varanasi city of Kashi where he turned the great Dharmacakra.
(d) The fourth stupa was contracted in the Jetavana of Sravasti where he performed the great miracles.
(e) The fifth stupa was constructed in the Kanyakubja city of Sankasya where the Buddha descended from the Trayastrimsah (the palace of thirty-three gods).
(f) The sixth stupa was constructed at Rajagrha where he reconciled the schism in Sangha (*sarighabheda*).
(g) The seventh stupa was constructed in the city of Vaishali where he prolongated his life span (*ayuhpramana*).
(h) The eighth stupa was constructed in the city of Kushinagara where he entered Nirvana.

Shapes and dimensions of the eight fundamental types of Tibeto-Buddhist *chorten*s are transformed types of earlier Indian Buddhist stupas. Despite this transformation, the Tibeto-Buddhist *chorten*s have retained all architectural aspects and religious significance as well as spiritual values and symbolical meanings which originated in ancient Indian Buddhist culture, as described in previous section.

Types of Chortens

*Chorten*s are of eight types. Structural distinctions amongst the types of *chorten*s are shown in Figures 3.6 to 3.13. These are explained below (Pema Dorjee[28] 1996) :-

(a) ***Chorten* of Enlightenment**. (Figure 3.6) It has a flight of four perfect square steps, which are free from embellishments. There is a tradition of making all the upper edges of the steps bulge out by one-quarter part of their own height to have an attractive look.

(b) ***Chorten* of 'Heaped Lotuses' Types**. (Figure 3.7) It is also called the *chorten* of "Auspicious Appearance" and "Appearance of Sugata". It comprises a flight of four circular steps adorned with heaped lotuses. There is also a tradition of making a maximum of seven steps to symbolize the seven paces that the Buddha took immediately after his birth in the Lumbini garden of Kapilvastu.

(c) ***Chorten* of 'Multiple Auspicious Doors'**. (Figure 3.8). It is also called the *chorten* of 'Divine Wisdom'. It comprises a flight of four square steps, and each side's centre bulges out by one-third the size of their width. These *chortens* have 108 doors or niches at the maximum, 56 in the medium and 16 in the minimum. Symbolically, four doors on each side symbolize the four Noble-Truths, eight doors on each side symbolize the eight doors of liberations, twelve doors on each side symbolize the twelve links of interdependent origination, and sixteen doors on each side symbolize the 16 types of emptiness.

(d) ***Chorten* of "Great Miracle'**. (Figure 3.9). This is also called *chorten* of "Conquest of Tirthikas', which is identical to the preceding ones (having a flight of four square steps), although it does not have doors.

(e) ***Chorten* of 'Descent from Heaven'**. Figure 3.10). Alternatively it is called "Offered by Devdas". The structural formation is similar to the preceding ones, but it has three ladders in the centre of each side's projected area.

(f) ***Chorten* of 'Reconciliation'**. (Figure 3.11). It comprises four octagonal steps formed by cutting off the four corners of the flight of four steps to form eight even sides. They are free from all embellishments. Occasionally, though rarely, there is a tradition of making all the upper edges of the steps projected out by one third of their own height.

(g) ***Chorten* of 'Victory'**. (Figure 3.12). It has three circular steps rather than four. Even the base of ten virtues and the vase-base are circular.

(h) ***Chorten* of 'Nirvana'**. (Figure 3.13). It is rather unique type of a *chorten* because of its 'bell shaped dome', in which mouth rests directly on the 'base of ten virtues' without any steps.

Structural Components of *Chortens*

The '*chortens*' in all its styles comprises three main structural bodies (See the figures on *Chorten* types):-

(a) **The Lion Throne.** This consists of six symbolic structures from the ground plinth upto the frame.

(b) **The Intermediate section**. This comprises six symbolical structures from the 'base of ten virtues' upto the *harmika*.

(c) **The Upper section.** The upper section or 'upper extremity' consists of all structures from the 'lotus supporting umbrella' upto the spherical pinnacle, and surround the 'axle pole', which is the vertical at the centre of a *chorten*.

Symbolical Meanings of *Chorten* Components

The flight of four steps, vase-base, the Vase (dome) *harmika*, axle poles, thirteen wheels and the rain-cloak comprise the physical structure of the *Chorten*. 'The accumulation' or 'essence' of *chorten* as explained by Angarika Govinda and Pema Dorjee is explained in succeeding paragraphs.

The Steps

The first step symbolizes the four close mindfulness (*catvari smrtyupasthana adhara vedi*) viz :-

(a) The close mindfulness of body.
(b) The close mindfulness of feeling.
(c) The close mindfulness of mind.
(d) The close mindfulness of wisdom.

The second step symbolizes the four perfect abandonments viz :-

(a) Efforts to abandon the non-virtues that have arisen.
(b) Efforts to prevent the non-virtues that have not yet arisen.
(c) Efforts to produce the virtues that have not yet arisen.
(d) Efforts to retain the virtues that have already arisen.

The third step symbolizes the four stages of miraculous powers (*catyari rdihipadas trtiya vedi*) viz :-

(a) A miraculous power possessed of compositional factor renouncing an aspiration of meditative concentration of meditative concentration or desire.
(b) A miraculous power possessed of compositional factor renouncing a meditative concentration of mind.
(c) A miraculous power possessed of compositional factor renouncing a meditative concentration of effort.
(d) A miraculous power possessed of compositional factor renouncing a meditative concentration of analysis.

The fourth step symbolizes the five (moral) faculties viz :-

(a) Moral faculty of faith.
(b) Moral faculty of effort.
(c) Moral faculty of maindfulness.
(d) Moral faculty of meditative çoncentration.
(e) Moral faculty of wisdom.

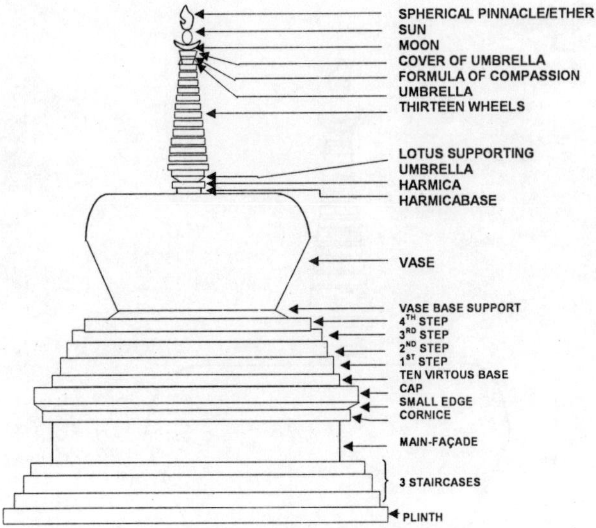

Fig. 3.6 Chorten of Enlightment

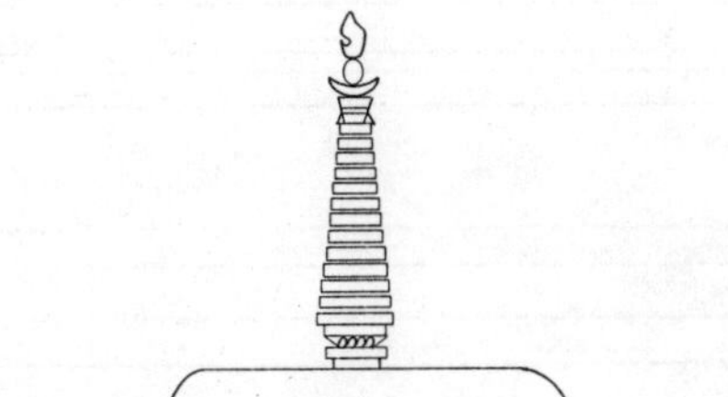

Fig. 3.7 Chorten of 'Heaped Lotuses'

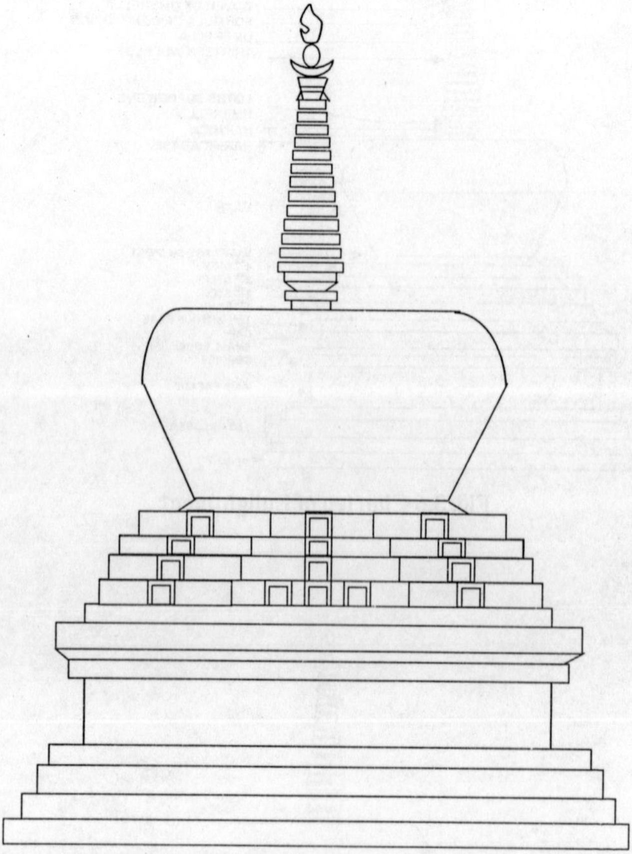

Fig. 3.8 Chorten of Multiple Auspicious Doors

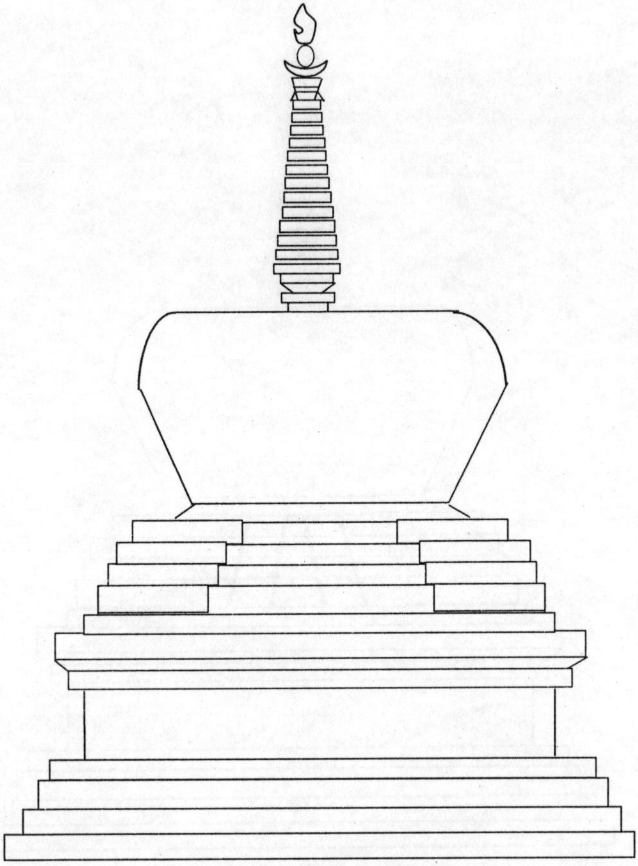

Fig. 3.9 Chorten of Great Miracle

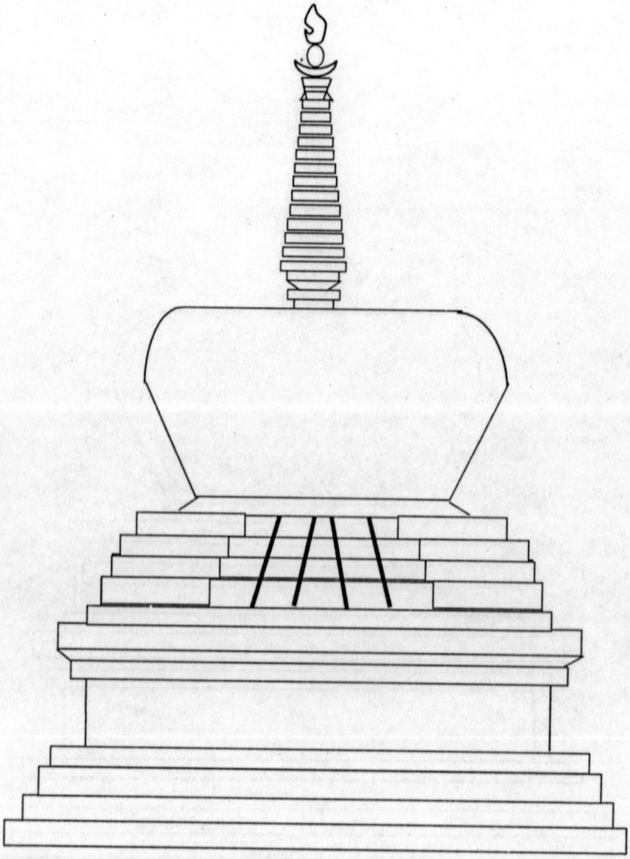

Fig. 3.10 Chorten of Descent From Heaven.

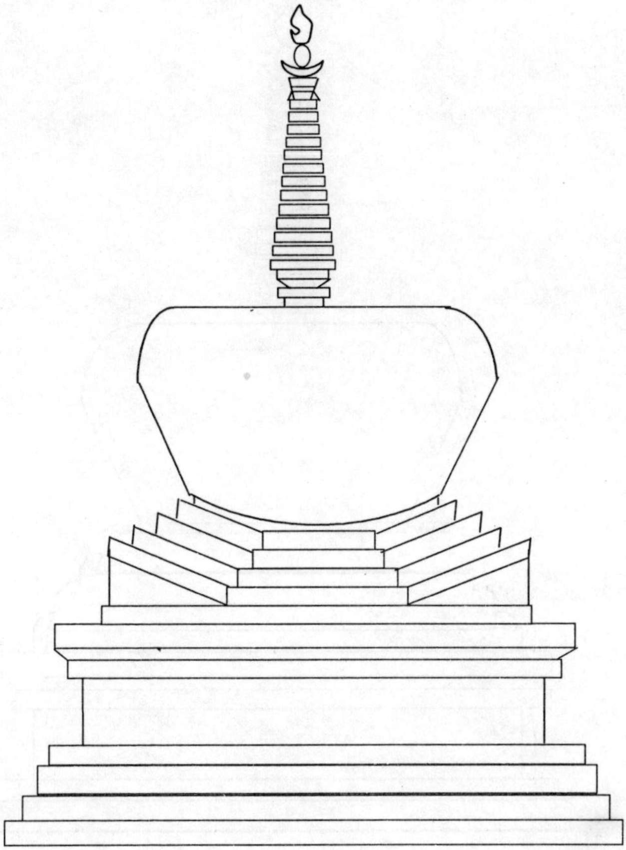

Fig. 3.11 Chorten of Reconciliation

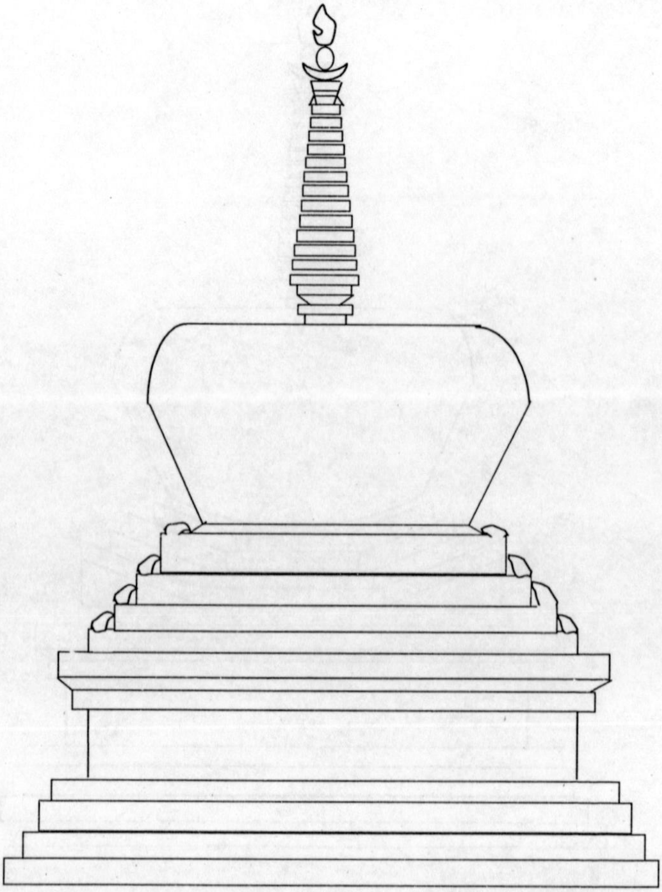

Fig. 3.12 Chorten of Victory

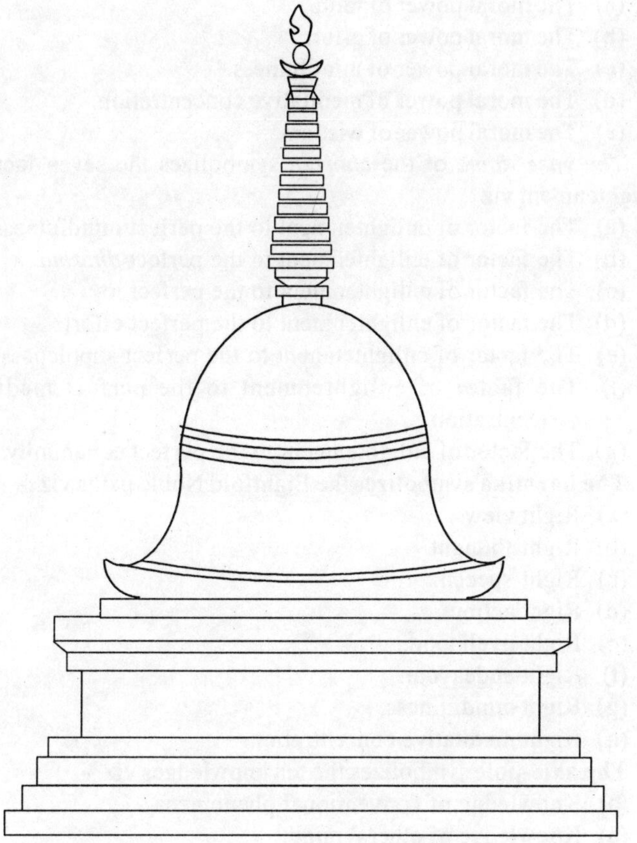

Fig. 3.13 Chorten of Nirvana

The vase

The vase - base symbolizes the five (moral) powers viz :-

(a) The moral power of faith.
(b) The moral power of effort.
(c) The moral power of mindfulness.
(d) The moral power of meditative concentration.
(e) The moral power of wisdom.

The vase dome of the *chorten* symbolizes the seven factors of enlightenment viz :-

(a) The factor of enlightenment to the perfect mindfulness.
(b) The factor of enlightenment to the perfect *dharma*.
(c) The factor of enlightenment to the perfect joy.
(d) The factor of enlightenment to the perfect effort.
(e) The factor of enlightenment to the perfect suppleness.
(f) The factor of enlightenment to the perfect meditative concentration.
(g) The factor of enlightenment to the perfect equanimity.

The harmika symbolizes the Eightfold Noble paths viz :-

(a) Right view.
(b) Right thought.
(c) Right speech.
(d) Right action.
(e) Right livelihood.
(f) Right endeavour.
(g) Right mindfulness.
(h) Right meditative concentration.

The axle-pole symbolizes the ten knowledges viz :-

(a) Knowledge of conventional phenomena.
(a) Knowledge of others' mind.
(b) Knowledge of *dharma*.
(c) Knowledge of realisation of subsequent.
(d) Knowledge of sufferings.
(e) Knowledge of cessation.
(f) Knowledge of sources of origination.
(g) Knowledge of paths.
(h) Knowledge of exhaustion.
(i) Knowledge of non-productivity.

Symbolism of the wheels is as explained :-

(a) The first bottom wheel symbolizes the power of understanding what is appropriate and inappropriate.

(b) The second wheel symbolizes the power of understanding that one is responsible for one's own deeds or action.

(c) The third wheel symbolizes the power of understanding, concentration, liberation, meditative stabilisation and meditative absorption.

(d) The fourth wheel symbolizes the power of understanding the superior and inferior faculties.

(e) The fifth wheel symbolizes the power of understanding the various mental inclinations.

(f) The sixth wheel symbolizes the power of understanding various mental faculties/constitutions.

(g) The seventh wheel symbolizes the power of understanding the paths leading to all goals.

(h) The eighth wheel symbolizes the power of understanding the recollection of former existence.

(j) The ninth wheel symbolizes the power understanding deaths and births.

(k) The tenth wheel symbolizes the power of understanding the exhaustion/cessation of contaminations.

(l) The eleventh wheel symbolizes the first peculiar close mindfulness(of the *Tathagata*), that by which, when the *Tathagata* teaches the *dharma* to his retinue, he does not get pleased when all his disciples listen with great respect.

(m) The twelfth wheel symbolizes the second peculiar close mindfulness, that by which, when the *Tathagata* teaches the *dharma* to his retinue, he does not get angry when all his disciples do not listen with due respect.

(n) The thirteenth wheel symbolizes the third peculiar close mindfulness, that by which, when the *Tathagata* teaches the *dharma* to his retinue, he neither becomes joyful nor angry when part of his disciples listen and part do not listen with respect.

The rain-clock symbolizes the great compassion by which the *Tathagata* examines the mental disposition of his trainees who are mature, immature or maturing. For the benefit of universal people, a perfect apprehension of all the truth bodies (*dharmakaya*) at all the different occasions is the nature of the stupa.

The umbrella represents the *Tathagata* a leader of all sentient beings, because the *Tathagata* became the supreme among the living beings without legs, among those having two legs, many legs, and among those who are corporeal and incorporeal.

The moon represents the brightness of the unsurpassed wisdom by which the *Tathagata* generates the eyes of the unsurpassed knowledges in the darkness of this world. Representation aspects of the sun and ether are given in previous section.

Chortens of Tashiding

Plate V of Figure 3.5 has been briefly touched upon in the previous section, where the description of Tashiding Monastery has been enumerated.

There are a total of 41 *chorten*s at Tashiding. They have been numbered for explanatory notes. Each number in plate V represents the location of the *chorten*s. The *chorten*s here are of three types:-

(a) ***Chortens* of Enlightenment**. Numbers 1 to 4, 8 to 17, 19 to 32, 34 to 36 and 38 and 39 Total-33 *chorten*s.
(b) ***Chortens* of Reconciliation**. Number 5 and 20.
(c) ***Chortens* of Great Miracle**. Numbers 7, 18 and 33.

*Chorten*s mentioned below correspond to the *Tathagatas* and *Gurus* as mentioned below :-

(a) **Dhudal *Chorten***. 8. This is the protector deity of Tashiding Monastery.
(b) Lhabab *chorten* is at No 24.
(c) Woe-Sung Ku-dhung, the oldest and largest *chorten* is at No 25.
(d) Kagyu Lama *Chorten* is at No 31.
(e) Lamyang Khentse Wongpo (Choe-ki-Lodoe from Tibet) is at No 23.
(f) *Chorten* of Guru Rang Jong.
(g) Dhakar (Write Rock) Tashiding, which is the key centre of main Tashiding pilgrimage is at No 40.
(h) No 41 is the location where Guru Rinpoche's arrow had contacted the ground, and where he sat in meditation.

Conclusion

Here we have reached the limit of what words can express. Beyond the words lies the practical relativity between the psychic, spiritual and natural forces, and this is exactly what the *chorten* of Tashiding, and other places in Himalayas and trans- Himalayas communicate to the living humans.

References

1. Jordan, T.G., and Rowntree, L., *The Human Mosaic: A Thematic Introduction to Cultural Geography*, Harper Collin Publishers, New York, 1990.
2. Durkheim, E., The Dualism in Human Nature in '*Essays on Sociology and Philosophy*', (ed). K.H. Wolff, New York, Harper Torch Book, 1964.
3. Swanson Guy, E, *The Birth of the Gods: The Origin of the Primitive Beliefs*, Ann Arbor, University of Michigan press, 1960.
4. Egri, P.C., 'Nature in Spiritual Traditions' in *Living with Nature* by Fischer, F., and Hajer, A., Oxford University Press, New York, 1999.
5. Toffler, A., *The Third Wave*, William Collins Sons & Co. Ltd., London, 1980.
6. Frost, P.J., and Egri, C.P., "The Shamanic Perspective and Organisational Change and Development, '*Journal of Organisational Change Management*, 71:7-23, 1994.
7. Halifax, J., 'The Third Body : Buddhism, Shamanism and Deep Ecology, 'in A.H. Badiner (ed), *Dharma Gaia : A Harvest of Essays in Buddhism and Ecology*, Calif : Parallax Press, Berkley, 1990.
8. Risley, H.H., (ed), *The Gazetteer of Sikkim*, Bengal Government Secretariat Calcutta, 1894.
9. Subba, T.B., *Politics of Culture*, Orient Longman Ltd, Chennai, 1999.
10. Foning, A.R., Lepcha : *My Vanishing Tribe*, Sterling Publishers Private Limited, New Delhi, 1987.
12. Macy, J., Toward a Healing of Self and World', in *Ecology* (ed) Merchant, C, Rawat Publications, Jaipur, 1996.
13. Chapple, C.K., 'Hindu Environmentalism: Traditional and Contemporary. Resources', in M.E. Tucker and J.A. Grim (eds) '*World Views Ecology*' Bucknell University Press, Lewisburg, 1993.
14. Gadgil, M. and Guha, R., '*This Fissured Land : An Ecological History of India*', University of California, Berkley, 1992.
15. Badiner, A.H., (ed), *Dharma Gaia : A Harvest of Essays in Buddhism and Ecology*, Galif : Parallax Press, Berkley, 1990.
16. Kaza, S., 'Acting with Compassion : Buddism, Feminism and Environmental Crisis', in C.J. Adams (ed) '*Ecofeminism and the Sacred*', Continuum, New York, 1993.
17. Bhatt, R., 'A Gandhian Perspective in the Himalayan Foothills', in J. Plant (ed) *Healing the Wounds*, New Society Publishers, Santa Cruz, 1989.
18. Devall, B., 'Ecocentric Sangha, in A.H. Badiner (ed) *Dharma Gaia : A Harvest of Essays in Buddhism and ecology*, Parallax Press, Berkley, 1990.
19. Risley, H.H., *ibid*.
20. GRIT Library, Wall Display, Gangtok, 2000.
21. Namgyal, Thutob, and Dolms, Y., *History of Sikkim*, 1908.

22. Dhamala, R.R., *Monasteries of Sikkim*, Occasional Paper No. 8, University of North Bengal, Siliguri, 1993.
23. Govinda, Lama, Angarika., The Historical and symbolical Origin of Chorten, in *Bulletin of Tibetology*, SRIT, Gangtok, Vol. VII, No. 3, 1970.
24. Yeshe, L., and Rinpoche, Z., *Wisdom Energy : Basic Buddhist Teachings*, Wisdom Publications, 1976.
25. Govinda, Lama, Angarika, *Foundations of Tibetan Mysticism*, B. I. Publications, New Delhi, 1960.
26. Dorjee, Pema., *Stupa and its Technology*, IGNCA with Motilal Banarsidass Publishers, New Delhi, 1996.
27. Govinda, Lama, Angarika,. *ibid.*
28. Dorjee, Pema,. *Ibid.*

4

Upper Rangit Basin : Aspects of Cultural Ecology

SECTION-I
INTRODUCTION

Geographical Setting

The geographical setting has played a major role in demarcation of the boundaries of nations and states and the cultural regions within them. The geographical setting has also had its profound impact on historical impulses and human ecology of the residents within a cultural region. The geography of Sikkim lends itself to one major cultural region, with three sub-cultural regions within it boundaries. While the Dongkiala and Singalila ranges demarcate the Eastern and Western limits of the State, the Chorten Nyima range in the trans- Himalayan region demarcates the Northern boundary.

The traditional trade routes from Tibetan Plateau into Sikkim were either through the Lachen or Lachung valley in the North Sikkim, or through the Nathu or Jelep passes on the Dongkiala range in East Sikkim. The Kanchendzonga Massif was a major geographical barrier leading to commercial isolation of Southern and Western districts of Sikkim. The Upper Rangit Basin in particular, wedged between Kanchendzonga Massif, Singalila range and Mongbrue watershed supported almost an independent sub-cultural region. Thus, the commercial activities, communication path ways, and altitudenal elevations divided Sikkim into following three cultural sub-regions :-

(a) Northern Sikkim sub region North of Mangan where the sparse population adapted to harsh and cold Tibetan type of climatic conditions.

(b) East Sikkim, which became a commercial trade link between Tibet and main land India.

(c) South and West Sikkim, where agro-economy and spirituality were more pronounced. Perhaps its geographical isolation presented itself as secure from political and military interference from outside, prompting the first Chogyals to establish their capitals at Yuksom and later Rabdantse in vicinity of the present township of Gyalshing.

Historically, Upper Rangit Basin had boundaries with the Limbuana country to the West, and Mongbrue watershed to the East. Culturally, it formed a part of Limbuana (Subba[1] 1999). Accordingly Limbus, Rais and Gurungs form a major part of population in Rathongchu watershed. The Bhutias, who arrived later, became the ruling class, the Chogyals being Bhutias themselves. The other Nepali castes, and Lepchas formed the remainder population in the area. The Lepchas, who are regarded as the autochtones of Sikkim, were animists and followers of Mon faith prior to their amalgamation into Mahayana Buddhist faith on arrival of the Bhutias. With the passage of time, cultural diffusion gathered momentum. The cultural diffusion has resulted in :-

(a) Though the Bhutia, Lepcha and Limbu tribes have their own languages, Nepali has become the *Lingua franca* in the regions.

(b) The Hindu temples have begun to hoist the prayer flags, which are original to the Tibetans' Buddhist cultural form. Many gods and goddesses in the Hindu Temples and Buddhist *Gompas* are common (Sakya[2] 1989)

(c) Names of all natural features, and human habitats originate from languages of all constituents of the population, mainly forming the Lepcha base. (Thakur[3] 1988).

(d) Ideas were borrowed from each other for rituals during shamans, and celebrations systems of festivals.(Subba[4] 1999).

(e) Methodology of food preparation and consumption patterns were learnt from each other, thus modifying the earlier practices of respective ethnic groups.(Tamang[5] 1998).

(f) The Nepali immigrants, barring the original Limbus, settled collectively as Nepali groups. This is in contrast to the Nepali demographic profile in mainland Nepal, where the various sub-groups such as Gurungs, Sherpas, Tamangs, Chettris etc are settled in their respective traditional geographic areas. This has resulted in a mixed Nepali culture, wherein cultural inputs have been borrowed from all constituent sub-groups. Intermarriages have further amalgamated the cultural mix.[6]

In the year 1788-89 Nepali Gurkha General Jahar Singh had invaded Sikkim, and captured areas up to the Western banks of Tista river. The area was under Nepali control for 29 years upto 1817, when the Titalaya treaty was signed. The Chogyal of Sikkim Tenzing Namgyal was forced to abandon his capital Rabdantse and retreated to safety beyond the eastern banks of river Tista at Tamlong. The captured territories, less the Darjeeling hills sector south of Rangit river were restored to the kingdom of Sikkim, while Darjeeling was merged with British India for good. J.C. White, the political Officer chose Gangtok for constructing his residence, and at his instance the Chogyal shifted the capital from Tamlong to Gangtok during 1885-86.

The restoration of lost territories of the Rangit Basin however did not prompt the Chogyal of Sikkim Tsugphud Namgyal to return to his original capital. The new capital city of Gangtok had now come into being and it continues to be the state capital till date . Eastern Sikkim and Kalimpong divisions had enjoyed the historical and traditional commercial bliss till the year 1962, when the Indo-China conflict put a stop to the trade links. The area however housed the capital city of Gangtok, and commercial centres of Rangpo and Singtam , resulting into continued economic, educational and urban development. The national Highway 31 A, which passes through Eastern Sikkim, helped the district authorities in setting up infrastructure for industrial development.

The West and South districts however prospered in spiritual and cultural ecology in isolation. The Upper Rangit Basin, being located in the remote salient, remained untouched by the popular culture. The ethnic and folk culture thrives in this pocket. The present day ethnic and folk culture is pronounced by:-

(a) **Folklores.** These include tales, songs, lores, beliefs, superstitions and customs that have been passed from generation to generation. The folklore and faith regarding the wishing lake Khechopalri deserves a detailed account. A detailed account of the same is given in a separate section in the chapter on spiritual ecology.

(b) **Shamans and Spirit Possession.** Shamans, *Jhakris* and spirit possession are encountered in all parts of the world. The Himalayas and other hill regions however have more cases of these kinds. Small case studies of various types of Shamans encountered in the study area have been analysed and described in later sections of this chapter.

Cultural Fusion and Identity

For centuries Sikkim continued to maintain its identity as an independent state under the Chogyals. The isolation of the kingdom helped the prospering of culture and spiritual related aspects within and outside the monasteries. Though the population percentage of Nepalese, most of whom are Hindu by religion is larger, the culture of the Mahayana Buddhist Lamaistic faith continued to be the state culture. The merger of Sikkim with India has its direct and indirect affects on the Sikkimese population.

The Sikkimese of today reflect the fusion of three different cultures of respective ethnic groups of Bhutias, Lepchas and Nepalese in varying proportions creating a kind of united Sikkimese culture. Additionally three new cultural entities, namely, traditional Sikkimese, Indian and Western have now taken roots. Their proportions depend upon the economic status; level of education and exposure to the world at large. The people in high income and higher education bracket may appear highly Westernized in external behaviour but they are very much conservative at heart. They go for Western education but strictly adhere to their religious and ritualistic practices at home. The mass participation of them in Sikkimese festivals, such as, Losoong, Bum Chu, Pang Lhabsol, etc., amply exemplify their adherences to the traditional culture. The upper class Sikkimese hardly show any interest in main land Indian culture.[7]

The middle class Sikkimese are comparatively less conservative. They have accepted both Indian and Western cultures alongwith their virtues and vices. They are highly influenced by the glitz that reach them via audio-visual devices and glossy magazines. The influences are manifest in their speech, attire and home décor. English being the medium of instruction in educational institutions, the young generation is leaning towards Anglicized life style.[8]

The people who are least educated and have a low economic status are understandably least exposed to the influences - Indian or Western, whatever the case may be. Even when there are less reservations, normally they follow their age-old traditions and practice whatever they have learnt from their ancestors. They are often found to adopt the food habits, clothings, and even means of recreation brought to them by the village traders who are mostly from the plains. The areas where everything from salt to sari come from India, are bound to have the influence of Indian culture.[9]

However, the villages located at the strongholds of the Kazi landlords have preserved their traditional Sikkimese culture under the strict vigilance of the landlords or the priests. Very recently, there was uproar over the construction of the Rathong Chu Hydel Project in Upper Rangit Basin at Yuksom on the issue of encroachment upon religious sanctity and culture. The protestations were so vehement that the project was scrapped mid-way. There are ample grounds to believe that the project has been scrapped to ward off the plainsmen who were supposed to construct the power plant, and in the process would pave the way for Indianization.

SECTION-II
CONSERVATION THROUGH CULTURAL LEGACY AND TRADITIONAL ECOLOGICAL KNOWLEDGE: KHECHEOPALRI LAKE–CASE STUDY

Sustenance of Ecosystems

Ecosystems sustain themselves in a dynamic balance based on cycles and fluctuations, which are nonlinear processes. Present day science has facilitated the study of this balancing pattern through field and laboratory research methodologies. Ecological awareness however only arises when we combine our rational knowledge with the traditional folk knowledge of the indigenous people. Such intuitive wisdom is characteristic of traditional cultures, such as the Sikkimese ethnic groups, in which life was organised around a highly refined awareness of environment.

The Sikkimese of Upper Rangit Basin residing in the 'spiritual and cultural pentagon' formed by the five monasteries of Dubdi, Khecheopalri, Pemayangtse, Tashiding and Ralang have had a symbolic relationship between their traditional spiritual culture and the environment. This traditional ecological knowledge has been acquired over centuries of trials. In general it differs from current scientific ecological knowledge in a number of ways:- (Inglis[10] 1993) .

(a) Traditional Ecological Knowledge (TEK) is mainly qualitative as opposed to quantitative.
(b) TEK has an intuitive component (as opposed to being purely rational).
(c) TEK is holistic (as opposed to reductionist).
(d) In TEK mind and matter are considered together (as opposed to a separation of mind and matter).
(e) TEK is moral (as opposed to supposedly value-free).

(f) TEK is spiritual (as opposed to mechanistic).
(g) TEK is based on empirical observations and accumulation of facts by trial-and - error (as opposed to experimentation and systematic deliberations for accumulation of facts).
(h) TEK is based on data generated by resource users themselves (as opposed to that by a specialised cadre of researchers).
(j) TEK is based on diachronic data i.e., long time -series on information on one locality (as opposed to synchronic data, i.e, short time-series over a large area)

In contrast to scientific ecology, TEK does not aim to control nature. The traditional people merge with the nature. The TEK is markedly slower than scientific knowledge, as it is based on human experience rather than scientific knowledge of tool and machines based experiments. The arrival of scientific development has to some extent, already shown signs of environmental degradation in this region. The modern housing techniques, the hydel projects, the electric power etc, have accelerated both-human development, and natures, and cultures' disintegration. The construction of roads and stone quarries have resulted in frequent soil erosion and landslides. The electric power based saw mills have led to faster clearance of the forests. Though, in Rangit Basin the situation is still manageable, but the degeneration needs to be checked.

In social and cultural context, the TEK of the ethnic habitat in Upper Rangit Basin includes:-

(a) Symbolic meanings through history, place names, and spiritual relationships between natural objects. Most rivers, Lakes and mountain tops here have spiritual relationships, Mount Khanchendzongpa being the "Guardian Diety" and "Khecheopalri" being the wish fulfilling lake.
(b) A distinct cosmology of world view based on the 'wheel of Life' and 'Mandala' add on of their conceptualisation of the environment, as opposed to the Western science. (Govinda[11] 1960).
(c) The human and environmental relations of all ethnic groups is based on reciprocity, and obligations towards, both, com*mun*ity members and other beings. This in the wider sense is also the essence of 'deep ecology', which is essentially biocentric.
(b) The 'Four Noble Truths' and the 'Middle Path-Madhyamika' or 'Eight Fold Path' essentially seek understanding and provision of means to support the humans alongwith the balance of the biodiversity in the region.

APPLICATION OF CULTURE AND TEK ON KHECHEOPALRI LAKE: CELEBRATION OF BUM-CHU FESTIVAL (FROM 12 to 13 MAR 1998)

The Lake Region

Sikkim is a land of lakes situated at elevations ranging from tropical to alpine. The majority of its lake ecosystems are of glacial origin and are located in the North District, while the South District has the least number. All recreational or other activities such as boating, water sports, and irrigation, etc. are restricted in these lacustrine systems, as a result of which the lakes of Sikkim have fared differently from the other lakes in India. Most lakes in Sikkim are considered sacred and are revered due to ancient mythological connections. Lake Khecheopalri is one among them. Although very few studies have been undertaken in general in the lacustrine systems of Sikkim Himalaya, notable reports on the various aspects of Lake Khecheopalri have been published by Venu *et al*[12] (1990), Roy and Thapa[13] (1996,1997[14]), and Roy *et al*[15] (1998).

Like other lakes in Sikkim, Lake Khecheopalri has multinomial names among the different ethnic groups of the state. Located within 27°19'15" N latitude and 88°15'06" E longitude at elevation of 1831m, near Tsozo village in the West District, it has the outline profile of a human right-foot. A cursory observation shows the original area of the lake to have been much larger in the past, and to have been later reduced to about a third of the open water area by the gross encroachment of marginal vegetation and by eutrophication. The lake has two permanent inlets and one outlet . Besides ground sources , water is also fed into the lake through two temporary inlets from surrounding hill-slopes over the monsoon. The rock in the lake area belongs geologically to the Sikkim group of granite-gneciss, schist and phyllites (Raina and Srivastava,[16] 1981, Bose,[17] 1989). The photograph on ensuing page reflects the siltation of lake on the Northern side. The cleanliness of lake water is seen in other photo.

Melas and Festivals

Sikkim is a mystic land of extraordinary beauty where a series of festivals are held and these annual festivals are commonly called "*Melas*". Some of the remarkable melas are the Khecheopalri *mela*, Saramsa *mela*, Rishi *mela* and Tashiding *mela* and the like.

Khecheopalri Lake: Silted Area.

Khecheopalri Lake: Neat, Clean Water Top.

The Khecheopalri *mela* in the West Sikkim is perhaps the largest and most significant of its kind. It was held near the Khecheopalri lake monastery on the occasion of "*Maghe Purne*" from Mar 12 to 13, 1998 and attracted pilgrims from all over Sikkim. Not only Indians, but foreigners also attended it. It had a special gravity. The pilgrims worshipped the spiritual power and offered various food-material into the lake to appease God. They used the lake's water as a "*Prasad*". From an ecological point of view, wheather the Khecheopalri *mela* was also an important contributor to the environmental degradation or depletion of aquatic biodiversity of the lake, is being assessed.

Purpose of the Festival

The Bum Chu festival got initiated at Tashiding monastery which is about 24 km away from the Khecheopalri lake. The festival generally coincides with *Maghe Purne* every year. There was a holy water-pot inside the monastery filled with water to the brim while prayers continued overnight. In the early hours these pots were uncovered in the presence of the High Lama by the monks. If the water showed ripples, that year could prove to be bad for Sikkim and vice versa. Since, Khecheopalri lake is the gift of Goddess Tara, Jetsum Dolma (according to mythological view), pilgrims assume the lake to be presented by the pot and used the lake's water as "*Panchamrith*". They pray and offer different products so that the lake will not dry and if it so happened people are doomed.

Pilgrims

Khecheopalri is an ethereal and venerated lake situated amidst dense temperate forests, with no floating leaf on the its surface. It is rather unbelievable. People of Sikkim believe that Lord Shiva existed in solemn meditation in side the lake. They also believe that the level of lake water increased during the festival which originated from the Head of Lord Shiva. Bearing these feelings pilgrims pay their visits to this holy lake during festival period from different corners of Sikkim and even neighbouring district of West Bengal. They come here to forget their sufferings and return home with the blessing of God. People from all religions Nepalese, Buddhists Bhutias and Lepchas were seen during the occasion.

Description of Festival

The Bum Chu (*Bum* means pot ; *Chu* means water) festival was celebrated near Khecheopalri lake only for two days namely, from Mar 12 to 13, 1998. People from all over Sikkim and other parts of India including foreigners attended this spiritual festival.

There were about 32 make-shift shops and restaurants including some grocery shops. All the shops were constructed using bamboo and katus (*Castonopsis sp.*) levels. There were some hotels serving food-items, like; chicken rice, mutton rice alongwith hot *momos, thukpa* and *kinema* (fermented soybeans) (special local food). Liquor (both local and IMFL mode) were up for sale. Beside these, villagers sold their local edibles like, *Ghar tarul*, *Simbal tarul*, *Ban tarul* (local tubers) and sweet potato. Other vegetables were also being sold.

Most of the pilgrims were piously engaged in their *puja*. Many pilgrims were in ebriated condition. The shop keepers were busy with preparation of food -items. An unexpected accidental fire in one of the stalls was noticed. These the pilgrims activities on the first day when the pilgrims number was about 1000.

The second day of the festival was started by Lamas' prayers. Pilgrims reached by all modes of transport. The level of pilgrims increased and some state ministers visited lake accompanied by officials at about 12:30 pm. It was the peak level of pilgrims about 2000 in number.

Pilgrims activities at Prayer station on Wooden Ramp

People of Sikkim believe that the lake Khecheopalri has some divine power to release the people from distress. They appease God by offering some materials. Even the temple offerings in the form of flowers, incense sticks cast into the lake's water every day take its toll. It was seen that pilgrims offered different types of fruits, milk, raw rice, fried rice, mustered oil, dalda (dehydrogenated oil), biscuits, bread sugar and the like into the lake. And the amount of these religious materials was maximum during the festival.

Moreover, they released many bamboo-boats with burning dalda and khadas (sacred scarves). This is the recent phenomena by the pilgrims. Animals, especially dogs (ferals) were seen relishing the proteinous offerings. Many officials including police and military personnel were also observed tying khadas to the offerings. One scene, very peculiar, was that of some pilgrims offering these materials to the water body and others consuming lake's water at the same time for their good health at this spot. Some washed hands unnoticed by the pilgrims who consumed the water.

The last day of the festival was a very significant day. Lake

Khecheopalri, a renowned water-body of Sikkim has many mythological tales. A team of monks (lamas) reached the wooden ramp to prove the reality of the lake's myths.There was the Khechu-Rinpoche, about 8 years old (believed incarnation) and a Cherechi Rinpoche, about 70 years old among them. The old monk folded few khadas spelling some spiritual words and threw them into the lake. The water of the lake then started bubbling after a few minutes. This was astonishing. Hence it was assumed, that the reality of living God existed in the Khecheopalri lake. It was the early morning feature on the final day.

Activities of Avifauna during the festival:

Nine species of avifauna were observed during the festival. They were little grebe *Podiceps ruficollis*, common merganser-*Mergus merganser*,large cormorant-*Phalacrocorax carbo*; little cormorant - *Phalacrocorax niger*; common teal-*Anas crecca*; tufted duck - *Aythya fuligula*; white breasted water-hen-*Amaurornis phoemicurus*; moorchen-*Gallimula chorophy*; and crane brown-Amaurornis bicolor. All these avifauna gathered at the center of the lake early in the morning and then dispersed due to human activities

Stall Construction:

There were about 32 stalls constructed using bamboo and katus (Castonopsis sp) branches collected from nearby dense forest. Each stall used about 50 kg biomass (total of approximately 1.7 tons of biomass including fuelwood).The tree-branches also served as fuelwood.

Sanitation:

Pilgrims defecated where ever they could (especially in the South and North slopes surrounding the lake). This definitely did not match with the sanctity in which the lake is held by the pilgrims, due to the shortage of proper sanitation facilities.

Waste materials:

About 5 kg of waste paper, plastic and polythene bags was detected after the festival, on Mar 13 1998 by a weigh-balance. The number of pilgrims was about 3000 during the entire festival duration. Each pilgrim carried about 15 gm offering materials. Thus, it was assumed that about 45 kg of materials was deposited into the lake.

Discussion :

Sikkim,is nature's paradise in the lap of the Eastern Himalaya. Bum Chu is an annual festival held around Khecheopalri lake and it is perhaps the most significant annual festival of Sikkim. Undoubtedly, fairs have great importance in our social life. But the fact was that the pilgrims in the festival helped accelerate the ecological degradation of the lake in the following ways:-

1. **Outwash :** Outwash (mainly triggered by activities of pilgrims) that was produced around the periphery of the lake during festival was observed as a pollutant of the once crystal clear lake. It is being accelerated due heavy rainfall.
2. **Cattle-grazing:** This is also a detrimental factor polluting the lake's water through excreta and is a common phenomenon surrounding the lake.
3. **Religious offerings:** Addition of offerings to the lake reduces the water-quality as found through the observations and results. This is a constant phenomena, but the highest level was during the festival.
4. **Loss of biomass :** Most of the shop-owners used the tree loppings with leaves from the surroundings for temporary construction which affected the lake's environment indirectly and decimated the biodiversity.
5. **Sanitation and hygiene:** There were no proper toilet facilities. Subsequently pilgrims used the surrounding areas (especially, South and North-East corner) of the lake for human defecation purpose. The pollutants would percolate into the lake's water during the monsoon and alter the ecological balance of the lake. There is no provision of water-sports, but pilgrims throw several products into the lake's without being noticed which certainly disturbed the biodiversity of the lake.
6. **Water quality:** The pH of lake water showed signs of being more acidic in nature. The possible reason for this could be leachates generated by the decomposition of waste materials.

Pollution and Water Quality Degeneration

The concentration of dissolved oxygen decreased with increase in temperature found after the festival. This is probably by increase in biochemical reactions due to addition of offerings which consume dissolved oxygen at a rapid rate and its concentration tends to decrease just after the festival. This was an important indicator of pollution .

Low content of dissolved oxygen just after the festival reflected the presence of organic matter in water offered during festival.

Chlorides are readily soluble in water. Changes in chloride contents between pre and post-period of festival, may be related to salts generated from the surrounding area through the activities of pilgrims. This also was an indicator of pollution levels.

The concentration of iron and ammonia increased and the high concentration of these also support the pollution of water. Some planktons like *protococus* and *Tetraspora* also supported this.

Therefore, the impact of monitoring of lake water chemistry indicated that the offerings and other activities influenced the water quality by away of pollution loading of the lake Khecheopalri. Hence, the lake Khecheopalri is faced with slow and certain pollution.

Proposed Remedial Measures

1. Afforestation of areas degraded by forest fires in 1970s should be undertaken.
2. Instead of prayers being offered at the lake proper, pilgrims should be encouraged into conducting prayer ceremonies at the monastery.
3. The management plan for the lake should be implemented through the Khecheopalri Development Authority formed by local people.
4. Education on the importance of lake ecosystems should be initiated, especially for school children.
5. Construction of houses, slash & burn agricultural and grazing in the lake catchment should be stopped.
6. The festival committee for the annual Bumchu festival should be authorized to collect and administer funds for proper sanitation, and provision of drinking water, parking facilities and garbage bins.
7. Anthropogenic outwash and agriculture runoff into the lake should be checked, alongwith landuse practices which contribute to degradation.
8. With the help of local community, mechanical and manual checks should be applied to control invasion of the lake by weeds.

SECTION-III
SPIRIT POSSESSION, SHAMANS AND JHAKRIS IN UPPER RANGIT BASIN

Introduction

Spirit possession in Sikkim is not unique. Social researchers and anthropologists have encountered it throughout the world. This section of the chapter on cultural ecology is an attempt at understanding some of the sociological and cultural implications of spirit possession with in the setting of Upper Rangit Basin. Mystical experiences are common in the Himalayas. Most people believe in spirits, and therefore fear the possession by the spirits.

There is also the religious angle to spirit possession. The concept of Mahayana Buddhism within Sikkim is based on the Tibetan lamaistic rites and the *Mun* Lepcha animistic rituals. The Limbus and the other Hindu Nepali castes have their faith in the ancient Hindu Tantric order mixed with animistic rites. The diffusion of the Bhutias Lepchas, Limbus and the other Nepali groups have however given an almost uniformly common platform to the phenomenon of spirit possession. Cultural borrowing of shamans is common.

Human Ecology Angle

From the point of view of human ecology, there is a close relationship between spirit possession as an altered form of consciousness and parapsychology. In speaking of the social function of human behaviour and in situations, it is useful to adopt the distinction between 'Manifest' and 'latent' function. Manifest function to do with intent while latent function is frequently unintended. In often used linguistic and anthropological terminology, manifest function is 'emic' while latent function is 'etic'.

To those who believe in spirit possession it provides a manifest function of the causes and effects of illness and misfortune. Throughout Sikkim, spirit possession is cited as a source of discontent as well as a means for resolving discontent. Illness and misfortune are attributed to a variety of supernatural forces such as attacks by witches, sorcerers, forest divinities, spirits of deceased individual and angry gods or goddesses.

People frequently recognize that the human neglect is usually at the basis for supernatural discontent. For example, the failure to propitiate household or village gods and goddesses is frequently cited as the source of their anger, or as among the Limbu, human envy and jealousy leaves the individuals or family open to attacks by witches or evil spirits. Through human failure and neglect supernatural forces are unleashed and bring chaos disorder and disaster. These forces are combated, held in check or defeated by individuals who are capable of 'seeing' and identifying them. Such individuals receive their power primarily through successful encounters with supernatural forces in the past. Frequently, these encounters are violent as in the initial possession of the *jhakri*. Invariably, the contact is loaded with danger. The eventual control of this contact imbues the individual with power to diagnose, cure illness and rectify misfortunes that are caused by similar encounters. They become 'living metaphors', through controlled spirit possession, or contact with supernatural power.

The system of cause, effect and cure, is thus a circular and enclosed system of knowledge. The cause is spirit, the effect is spirit possession, and the cure is controlled spirit possession. This system of knowledge provides the manifest function of explanation and control in the face of disorder, chaos and inexplicable circumstances. The social function here described is adaptive. It (spirit possession) is the basis of 'social stability' in potentially unstable and disruptive social circumstances. It has the similar function that witchcraft beliefs, have for many societies. It functions to 'explain unfortunate events' and thus stabilize social relations where doubt serves to disrupt them.

Meanings and Definitions

Shaman : In some of the general dictionaries and anthropological text books, the term 'shaman' is defined so generally that no elements differentiate it from words referring to many other religious practitioners. For example in the *The Random House Dictionary of the English Language* (Stin1973:1310) '*shaman' is defined as a medicine man; one acting as both priest and doctor who works with the supernatural'* Keesing (1958) in a general anthropology textbook, defines 'shaman' as an individual religious expert' (Hitchcock and Jones 1994).[18]

Spirit Possession. The invasion and control of a person by a spirit, demon or god. Spirit possession is characterized by a change of personality and behavior on the part of the possessed. In some societies illness is regarded as a form of spirit possession.

In Sikkim the common name for all shamans in general is '*Jhakri*'.The shamans in the Sikkim com*mun*ity context are outlined as under:-

(a) ***Limbu Shamans***: Limbus recognize five kinds of shamans whose names are '*Yeba*', '*Yema*', '*phedangma*', '*samba*' and '*mangba*'. The differences are minute. All perform at weddings funerals, harvest rites and periodic ritiuals for the welfare of households. The differences are more pronounced when it comes to the diagnosis and cure of disease and the prevention of misfortunes and other calamities that have supernatural causes. '*Yeba*' is male while '*Yema*' is a female '*Samba*' is a specialist in oral literature and mythology. '*Mangba*' is a specialist in dealing with the spirits of those who died through violence.

(b) **Sunuwar**: '*Punimbo*' and '*ngiami*' are the shamans of Sunuwar com*mun*ity. Their numbers in the Upper Rangit Valley are very few though.

(c) **Rai**. The Thulung Rai com*mun*ity belonging to Tibeto-Burman (Kirata) sub group refer to their shamans as '*Saleme*', while Nepali speaking Rai com*mun*ity refer to them as '*bijuwa*'.

(d) **Sherpa**. Sherpa shamans are known as '*hal-wa*' meaning 'God one' or '*min-dung*'. The female shamans are called '*do lo ma*', neaning 'woman who has been to hell and returned back'.

(e) **Kami**. The Kami *jhakris* are most numerous. They are some times addressed as '*dange*'. The wife of a *jhakri* is called '*jhakrini*', while a *jhakri* is a female shaman.

(f) **Tamang**. Tamang '*jhakri' is called 'bon-po*'. The head dress of Tamang shamans, and other gear has a similarity with other '*jhakri*' groups.

(g) **Gurung.** There are two categories of Gurung shaman, the '*pajyu*' and the '*khepre*', Both traditions incorporate pre-Buddhist Bon rites . Aspects of Lamaism and Nepali Hinduism are both detected in ritual style and paraphernalia.

(h) **Lepcha.** The chief functionary of their indigenous beliefs is '*bon-thin*', who is a kind of intercessor. The female is '*mun*'.

Shaman and Caste System. Within Hindu com*mun*ity the caste system and ethnic pluralism operate to deny many individuals self-respect and prestige in the village setting. This is especially the case with lower castes such as metalworkers, tailors or leatherworkers. Frequently, these castes provide the village shaman or religious specialist who combats disease and misfortune through controlled spirit possession. In respect and prestige, spirit possession in all its forms is not limited to any particular class or caste. It is a sociologically 'open' system. This is not to say that it occurs with the same frequency among all classes and all castes, but simply that it is open to all. It is a recognized and frequently institutionalized means for social advancement (Field study[19]).

In this fashion, the beliefs and institutions surrounding spirit possession fulfills function, of providing 'a culturally approved means for the resolution of inner conflict (between personal desires and cultural norms)'. Within the Hindu caste system many people are denied access to goods, wealth and prestige as a result of culturally prescribed norms. The institution of spirit possession within different context serves as a means for the satisfaction of personal wants and desires. Thus on one level it provides an explanation for social dissatisfaction and on another it channels it according to prescribed norms.

Female Shamans. The spirit possession in general occurs more frequently among those categories of people who are denied social statuses and traditional means for social advancement in society. This is especially in the case with women, who more frequently than any other social category, find themselves relegated to inferior positions in society with little or no hope of social and institutional advancement. Women of lower castes and tribes would by definition be in relatively inferior social positions to women of higher castes who, through marriage and education, are in positions of respect. Unmarried women, widows and women in servants positions are more likely to become possessed by spirits than married women with children, a family and a successful marriage. Spirit possession, as pointed out by Lewis (1971), provides an oblique aggressive strategy' to such socially deprived categories of people.

Male Shamans. Similarly, men of lower castes, or men without lands, titles or positions of authority seem more inclined to spirit possession than those of higher castes, with wealth and social position. Spirit possession again provides an approved means for 'social protest'.

Distinction between Shaman and Priests. Shamans are generally distinguished from priests by the following attributes:-

(a) Shamans are part-time professionals and seldom make their living solely through their religious occupation, as does a priest.

(b) Shamans do not inherit their status or succeed to, but acquire it by divine intervention, which generally occurs in adolescence as a result of the trance or a state of possession. The priest, on the other hand, lacks the psychodrama of the shaman and generally acquires his position through inheritance, study or other channels of succession.

(c) Shamans are primarily curers and healers, while priests seldom diagnose disease and misfortune but instead limit their activities to prayer, worship and religious offerings to supernatural beings.

(d) Shamans generally shy away from set rituals having to do with the life cycle, planting, harvesting or temple worship, while these are the stock in trade of a priest.

(e) Shamans are controlled ecstatic who assume a trance or become possessed, and in that state they serve as mediums through which supernatural beings, make their wishes known, and display their power in a dramatic style.

Case Studies

Case studies on certain specific shamans, *jhakris*' and '*bon-thins*' encountered during field study are enumerated in succeeding sections of this chapter.

SECTION-IV

SHAMAN PRACTICES AMONGST LIMBUS

Introduction

Western district of Sikkim has been historically and culturally very closely associated with the Limbuana cultural centre of Nepal. Limbu consider themselves to be autochthones of the area. Limbu shamans are common in general area Yuksom, and areas West of Rathong Chu. Various kinds of Limbu shamans have been described in the preceding section of this chapter.

Limbu shamanism rests on firm belief that people's lives are intimately linked to world of supernatural beings. That world consists of numerous gods, goddesses, ghosts of the dead, and spirits associated with animals, plants, the forests, the home and other animate and inanimate objects of nature. Not every Limbu believes the same way or in the same things, nor as in most other cultures, does every Limbu believe in supernatural forces with the same intensity. I encountered many Limbus who were confirmed agnostics, while others could be classified as 'true believers'. In general, most Limbus accepted the idea that disease, misfortune, crop-failure, death and other unexplainable events had supernatural causes. In order to understand these causes and to rectify unfortunate circumstances, they seem to depend with profound faith on the shamans or '*jhakris*'.

Limbu Shamanic Costumes

The differences between Limbu shamans are also defined in their respective costumes and shamanic paraphernalia. Both the *yeba* and *yema* wear a costume consisting of skirt (*jama*), feather headdress (*wasang*), *cowrie* shell bandoliers (*aplak*), and garlands of acron seeds, *rudraksa* beads, bells, and the like. They frequently wear a leather or canvas belt (*peti*) on which are strung a series of bells. The *yeba* and *yema* do not use the traditional shamanic drum found throughout Nepal, but instead during a state of trance or possession they beat a brass plate (*thal*).

In contrast, the garb of the *phedangma* and *samba* is much less dramatic. Noticeably absent are the feather headdress and *cowrie* shell bandoliers. Only on a rare occasion, depending, on the teachings of one's *guru*, did I see a *phedangma* or *samba* wear the shaman's skirt (*jama*).The key instrument in the ritual activities of the *phedengma* and *samba* is the two headed drum, the *dhyanro*, and both wear garlands of *rudraksa* beads and bells in curing ceremonies. It is important to note that shamans of all types seldom use the drum or the brass plate in life-cycle rituals birth, marriage, and death-or harvest rituals. Their activities during such rites parallel those of a priest, while their powers as shamans are used as a means for summoning spirits and tutelary deities, as well as maintaining power over these supernatural beings. These powers are undesirable when making offering to gods and goddesses who are neither angry nor displeased. The *mangba* uses both the drum and the brass plate, but does not wear the feather headdress, skirt or *cowrie* shell bandolier.

The costume and instruments of the Limbu shamans indicate a diffusion of all shamanic practices and resemble those of other groups in Sikkim. This is especially evident in the garb of the *yeba*. During the course of field work, I attempted to witness and check a full costume of the *yeba* and *yema*. One Limbu informant took me to a Tamang Shaman (*bon-po*) who owned such a costume. He was an old man of about seventy and no longer practicing his art. The headdress was almost identical to the ones seen elsewhere, with the exception that the Tamang headdress (called by the same name, *wasang*) consisted of peacock feathers, while the *yeba*'s head dreses included the feathers of a number of other species of birds. Both headdress displayed a headband decorated with *cowrie* shells in which the feathers were sewn. During my search, I also encountered a female shaman (*yema*) who owned a similar headdress but refused to wear it . The type of headdress under discussion closely resembles those of the *jhakri* found elsewhere in Sikkim.

The *yeba* told me that the kinds of feathers used were not of extreme importance, although the wild species of birds were preferred to domestics fowls. The feathered headdress of the *yeba* is symbolic of celestial flight, characteristics of classic Nepali and Sikkimese shamanism, which explains the preference feathers of wild birds, especially wild birds of the high Himalaya.

Other instruments of the Limbu shamans parallel the common tradition. The two headed drum (*dhyanro*), oval, or occasionally diamond or triangular shaped, resembles many of the drums used by other shamans. Although the hides of goats and water buffalo are occasionally used in making drumheads. The most preferred skins are of deer and elk. Drum heads are occasionally painted with drawings of the sun and moon, which are symbolic of the shaman's celestial origin and flight during the trance. Similar motifs are frequently carved with images representing helping the Limbu high goddess Yuma Summang, the earthly counterpart of Tagera Ningwaphuma, the creator. A number of species of animals are carved on drum handles-birds, elk, deer, and snakes are the most frequent. Most of these animals are symbolic of his journeys through the sky during a state of ecstasy. Also, frequent representations of the 'cosmic tree' are either painted on the drum head or carved in the handle.

Drums are used by Limbu shamans to achieve a state of ecstasy and control helping spirits as well as to gain power over those that do harm to the living, or the spirits of those who died by violence. Although the *yeba* does not use the shamans drum during a seance, preferring the brass plate.

Divination and Curing

During course of field work I spent a day with a Limbu shaman Purna Bahadur. He explained his methodology of divination and curing during shamanic rites.

Limbu shamans divine in a number of ways. The techniques he uses are related to the different situations of illness and misfortune. The most frequent technique used for minor aches and pains is rice divination. The patient or client presents the shaman with a coin, which he places in a brass plate. He then takes a handful of rice, covers the coin, and separates the rice into six small piles. I was told that four of these represented the direction and two of them the Limbu deities. In one case, the shaman told me that two of the piles symbolized *Yuma* and *Theba Sammang*. The shaman then counts the grains of rice in each pile, even numbers signifying one thing and odd numbers another. In this way he is able to determine the severity of the illness, its cause, the direction from which it came, and the course of action to be taken.

A second form of divination, used in cases of extreme illness, requires the shaman to take articles of clothing from the patient and place them at an altar erected for *Yuma Samang* or other Limbu deities. The shaman then chants the *mundhum* and goes into a trance. He learns the nature of the illness by divination in the presence of the clothes and seeks help from his tutelary spirit. The details of this form of divination were never explained to me, but it is used in most major seances where the shaman constructs and altar and goes into a trance.

Occasionally he would take the pulse of a patient and determine the illness. At a wedding he attended, one of the guests fainted. Two shamans were present. One held the patient and took his pulse beat. After a few moments, he declared that the man had been attacked by *Tamphungma*, the jungle sprit. He chanted a few *mantra* to this spirit and gradually the patient revived. He told me this would not free the man from the grasp of *Tampungma*, but was only a temporary measure. On returning home, the man would have to call in shaman who would perform a *puja* to *Tamphungma* on his behalf.

As stated earlier, all illness and misfortune are thought to have supernatural causes. In extreme cases, the divination and cure require elaborate praparation on the part of the shaman and are expensive for the patient. The patient and his family must feed the shaman while he is in attendance. They should also supply him with liberal quantities of *chhang* liquor and tobacco. The payment of the shaman follows the cure. I was told that unsuccessful cures required no payment but that a successful cure would sometimes require as much as hundred or more rupees. The amount would be determined according to the time the shaman spent with the patient. Sometimes payment is in cash, but more frequently, it consists of goods such a clothing or grain.

Becoming a Shaman

One becomes a shaman as a result of possession by a deity.The most frequent possession seems to involve *Yuma Sammang*. At first, possession is involuntary, and the victim seeks a cure by consulting a shaman, who will attempt to pacify the deity with offerings. Should he be unsuccessful, the victim may take this as a sign that he is to learn the art of shamanism in order to eventually control his possession.

Many times, the state of involuntary possession continues for years, or in rare cases, throughout one's life, without the victim becoming a shaman. Simple possession by a deity is not the only sign that one is to become a shaman. One must acquire a teacher, or rather, more accurately, the teacher must discover the patient and learn the will of the deity. Limbu insist this knowledge comes to the teacher as a result of the deity's own wishes, not the patient's or the teacher's. The teacher might learn this in a dream, or in a state of trance, or through some other divine omen. He then pays the victim a visit, informing him of the wishes of the deity. The teacher who is himself a practicing shaman, then begins to teach the patient the *mundhum* and the techniques of his art, which include divination the construction of an altar, and how to make and use the shamanic paraphernalia-the costume, the drum and the various fetishes.

Invariably, the patient rejects the call to become a shaman. Few desire to become a shaman but, at least consciously, would rather be cured of their possession than to take on trade which requires much time and effort and few awards. I never met a wealthy Limbu shaman. The practicing shaman is frequently taxed both physically and mentally. During major festivals, such as Dasain, he is in great demand and often goes days without sleep. Frequently, his own family is neglected, as well as his fields. The re*mun*eration for his services seldom equals the losses that he suffers as result of neglecting his own agricultural work. Furthermore, most of the shamans I met had large families. One shaman had eight children. The value of a large family is great, once the children have reached the age where they are able to help in the field, but while they are young, they only represent more mouths to feed.

Spirit possession indicating the call to shamanic apprenticeship usually occurs at an early age. One shaman, a *yeba*, was first possessed at the age of eight by the spirit *Sandonogma Yejuli*, a household spirit of his mother's brother. His family called a shaman on a number of occasions, with the intention of curing the boy, without success. Finally a *yeba* learned of his plight and visited the boy, telling him that he was to become a *yeba*. He taught him the *mundhum* and how to control the *Yejuli* spirit. At first he rejected the idea, but after much coaxing from his *guru*, he decided that he would become an apprentice. His apprenticeship was short, not lasting more than two years. He then began to call the spirit at will and have success in curing people who became ill. Occasionally, however, he has difficulty in controlling *Yejuli*, and sometimes the deity, 'comes on him' without being summoned.

Yuma Sammang is frequently the goddess that possesses an individual and forces him or her to become a shaman, but it may also be spirit of one's mother's brother's patrilineage, or the household deity of one's mother's brother, which are believed to be female deities as well. It may also be the spirit of one's maternal grandfather or grandmother. There is a widespread idea that the tutelary spirit of Limbu shamans are 'inherited' in the matriline and are feminine.

Being called by a spirit, through possession, does not always mean that the individual will become a shaman. One man who lived near Yuksom would occasionally become possessed by the spirit of his maternal grandfather, but he did not become a shaman. I was told that a *guru* never came to him and he never learned the art. It was emphasized that , if the *guru* did not take up the individual as an apprentice, the individual could never learn to control his helping spirit.

The *guru* is essential to the practice of Limbu shamanism. He is the one who has learned to control his possession. He teaches this art to his apprentice and instils in him the confidence that is needed in dealing with supernatural powers. In this way, the *guru* also teaches his apprentice the causes of illness, the ways to diagnose, divination and the means of curing. Without a *guru*, a person possessed is merely a person that is ill.

Conclusion

Limbu shamanism is not just a matter of teaching and learning the procedures. Without the call through possession by a spirit, a person can never become a shaman, regardless of how much he learns about divination and curing rituals. Possession and teaching are both essential to the profession.

SECTION-V
THE LEPCHA - BON- THIN OF TASHIDING

Introduction

The Lepchas living in general area Tashiding adopted Tibetan Buddhism in 17th century. But this has by no means involved the elimination of their earlier religious traditions. The chief functionary of their indigenous beliefs is the *bon-thin* who is a kind of intercessor. He is chosen by the tutelary god *bon-thin-rum* from a family where there have been *bon-thin* for many generations, and he is initiated after a long illness which serves as a kind of preparatory phase. The occupations of the *bon-thin* are the same as those of other villagers. He cultivates his field and looks after his goats, cows and oxen. Because of his religious position, he possesses a high social standing in the village.

The well-intentioned divinities are called *rum* in contrast to the divinities of evil intent, the *mun*, who are the cause of disease, death and disaster. The most important of the propitious divinities is *ta-se tuk-bo-thin*, also called *ta-se-thin*. Many of the propitious divinities take the form of female spirits (*mit*) inhabiting springs and *chus*. The evil divinities, the *mun*, are innumerable, haunting forests, bogs, torrents and rocks. Being very powerful, they are liable to interfere at any time in the life of the Lepchas and obstruct their projects. Only repeated sacrifices can satisfy their greed for flesh and blood.

Case Study

A double religious function devolved upon *bon-thin* Phurbu who was my informant. In the role of propitiator he had to attract the attention of the chief god *ta-se-thin* and other well-intentioned divinities by means of offerings; as intercessor he had to appease and drive away the *mun* through blood sacrifices.

Periodically the *bon-thin* burns incense and recites the following prayers.

'O great god, protect us from illness, influenza, winds of the South and lightning. Protect us from the evil which can come from those who practise magic against us. You, too, king of demons, restrain your followers'.

Invoking Deities

The most important religious ceremony (*rum-fat*) takes place once a year after the harvest. The *bon-thin* makes an offering to all the divinities with whom he has contact. The villagers bring rice, millet, eggs, butter and fruit of all kinds. These offerings are placed on a stone on a hill-top well in view of *kon-cen cu* (Khanchandzonga), the mountain deity which dominates the whole region.

The *bon-thin* invokes successively the different summits of the mountain, which are identified with the following divinities : *kun-cen cu* and his wife *kon-lo-cu*, *pun-dim* and *pun-zon-cu*, their son and daughter *ra-tat* and *ra-yot*, their grand-son and grand daughter *sin-*

mu and *ko-hun-mu-cu*, their male and female servants. The offspring and the servants are the lesser summits on either side of *kon-cen cu*.

Next the *bon-thin* addresses the spirits of the earth and the protectors of the village places, *tsen-gog* a mountain divinity, *ta-lyan-do* divinity of the heights above the village, *so-mon-pon-di*, spirit of the hamlet. To these divinities, whose role seems altogether rather obscure, he offers rice, millet and oranges. In the third place the *bon-thin* requests the chief of the demons to restrain his followers. Of these *tam-nok-mun* and his partner *lyan-ser* were mentioned as the most powerful. This pair of black demons comes from the plains; having no children they are particularly self-willed and so more difficult to control than the others who are named *tsen-dut* and his mate, *tun-kun-mit*, *lun-li* and his mate *lun-lun-mun*, dwellers in caves, rocks and streams. The *mun* receives offering of *chhang* and rice.

Sacrifices

In spite of such periodical offerings and sacrifices of cocks, the *mun* still manifest themselves by causing diseases. The *bon-thin* is called upon in these cases to identify the dissatisfied *mun*. By means of divination he determines the kind of sacrifice required by the demon. A serious illness often necessitated a sacrifice of an ox. Before killing the animal invocations were made to the demons *gie-bu-mun, co-gye-mun, tsen-dut* and *me-so-mun*, who takes the life of men. While the animal was being killed, the sick man was attached to it by means of a cord, and he beseeched the demon to accept them, praying that they may cease to cause illness of all kinds: headache, pains in the limbs, etc. The sacrifices of the ox are on a wane these days.

Functions on Death

Finally the *bon-thin* has an essential function to play at the death of a man. Since the soul of a man must return to the place of its origin, it has to find its way to the ancestors *ti-kun-tik* and *ni-kun-nal* in the sacred land *rum-lyen*. In order to make the soul take the right path the *bon-thin* has at his disposal suitable spells. If a man is killed in the jungle or eaten by an animal, this causes special difficulties. Great efforts are made to retrieve his body, for if his soul has not been properly dispatched, it will hang around troubling members of the family. My informant also told me that the soul is judged by the god *co-ge pe-no* and is sent either to heaven or hell.

Mun

Mention was also made of the *mun*, the feminine counterpart of the *bon-thin*, but none is to be found now in Tashiding. The *mun* whose office passes from grandmother to granddaughter, is inspired by a tutelary divinity similar to *bon-thin rum*. According to the Tashiding villagers, these *mun* have power over the divinities of streams and mountains. Their supernatural power is characterized by the following stories which are often told in the village.

Folk Tales on '*Mun*'

Two brothers had become enemies after quarrelling about a piece of land. The younger one appealed to a *bon-thin* in order to avenge himself on his brother. The elder one fell ill soon afterwards, and so his family called upon a *mun*. By means of divination she discovered the cause of the illness. A pig had been buried in the field which was the cause of the dispute. This pig was still alive, and the *mun* ordered that it should be dug up and placed near the house of the younger brother. Four days later, he died, together with seven members of his family'.

In earlier times, the *mun* shared with the *bon-thin* the power of guiding the souls of the dead to *rum-lyan*. Moreover, they could forestall the future and hold conversation with the dead.

'Three Lepchas went fishing. One of them left his companions and disappeared. His friend looked for him and found his corpse in the river. On returning to the village, the two men, although innocent, were accused of his death. A *mun*, who was consulted, asked the soul of the dead man how he had died. He replied that a *mun*, whom he had offended by cutting bamboos in a forbidden place, had pushed him into the water. Through the mediation of the *mun*, the drowned man asked his father to release his friends and also revealed the place where his savings were hidden. The parents found in fact eleven rupees and some paise in a bamboo near the hearth. The dead man entrusted his child to his parents and requested his wife not to marry again'.

Conclusion

The '*bon-thin*' and the '*mun*' are different from the other '*jhakris*' in the sense that, these also carry out the tasks of the priest hood. Though, since the merger of Lepchas within Mahayana faith, the dependence on the on the lamas and the monks has become more pronounced, the traditions of the '*bon-thin*' and '*mun*' continue.

SECTION-VI
FOLKLORE RELATING TO JHAKRIS

Introduction

The word folklore –literally, "the learning of the people"-was coined in 1864 by the late Mr. W.J. Thomas to replace the earlier expression "popular antiquities". It has established itself as the generic term under which the traditional beliefs, customs, stories, songs and sayings current among traditional people or retained by rural ethnic classes of more advanced peoples. It comprises nearly all animistic beliefs about the world of nature, animate and inanimate; about human nature and things made by man; about a spirit world and man's relations with it; about witch craft, spells, charms, amulets, luck omens, disease, and death. It further includes customs and rites at a marriage and inheritance, childhood and adult life, and as to festivals, warfare, hunting, fishing, cattle keeping etc; also myths, legends, folk tales, ballads, songs, proverbs, riddles, and nursery rhymes. In short, it covers everything. (Burne[20] 1914).

The subjects comprehended under the name of folklore may be arranged in three principal groups with sub headings as follows (Burne[21] 1914) :-

(1) **Belief and Practice Relating to** :-

- (a) The Earth and the sky.
- (b) The Vegetable World.
- (c) The Animal World.
- (d) Human Beings.
- (e) Things made by Man.
- (f) The soul and Another Life.
- (g) Superhuman Beings, (Gods, Godlings, and Others.)
- (h) Omens and Divination.
- (i) The Magic Art.
- (j) Disease and Leechcraft.

(2) **Customs.**

- (a) Social and Political Institutions.
- (b) Rites of Individual Life.
- (c) Occupations and Industries.
- (d) Calendar Fasts and Festivals.
- (e) Games, Sports, and Pastimes.

(3) **Stories, Songs and Sayings.**

- (a) Stories: (i) told as true; (ii) told for amusement.
- (b) Songs and Ballads.

(c) Proverbs and Riddles.
(d) Proverbial Rhymes and Local Sayings.

Sikkimese Folklore

Stocks[22] (1926) and Kotturan[23] (1989) have given detailed accounts of Sikkimese folklore, most of which are, at least in numbers, dominated by Lepcha folktales. *'Jhakris'*, shamans and *'bon-thins'* form an integral subject within the arena of folklore and customs. Almost all subjects described in introductory paragraph above form the vast assemblage of folktales.

The folk tale described here is conspicuous by the fact, that it encompasses within itself the Hindu Gods, and the Buddhist Bhutia Lama. The tale also to a great extent explains the diffusion of the shamanic rites within the region. It relates to the origin of the *'jhakri'* system, and as to, who was the first *'jhakri'*. The ethics to be maintained by the 'jhakris' are also amply brought out in the tale.

Folktale of First Jhakri

Majority of persons agree that Shankar Mahadev Bhagwan was the first 'jhakri'. Mahadev and a Bhutia lama met on Mount Kailash in Himalayas. The two competed to draw forth from the jungle, using nothing but 'Mantra' the following objects:-

(a) ***A 'ghorin'.*** A necklace made of beads. This type of necklace is also used in Tantric ceremonies. Each 'jhakri' possesses several of these.

(b) A *'rudraksha Mala'*.

(c) ***Dumsi ko karo.*** These are the porcupine quills.

By the same 'mantra' power they also tried to fill an empty vessel with water and reach the sun. Mahadev produced a *'ghoirin'* and *'mala'*. Mahadev filled the vessel with water. The Lama drew a prayer wheel. Mahadev arrived half way to the sun, but everything become dark, and he could no longer follow the path. The Lama arrived at the sun. Mahadev constructed a drum and danced and played with it. Since Lama had reached the sun, it was agreed that Lama would perform *'puja'* during day, whereas Mahadev would perform at night. Mahadev and Lama then separated and each settled down in different parts of Kailash.

The story continues on making of Mahadev's drum. Mahadev sent the porcupine into jungle to bring *'sinaro'* wood from the jungle to make the drum frames. A deer skin was brought for making drum heads. Mahadev closed the drum by attaching a handle. A *'rudraksha mala'* and other articles were placed inside the drum. A hen was brought which laid an egg.

A red dog arrived by itself. The dog irritated the Lama, who killed it. The dog's spirit assumed the role of guardian of the drum and the egg. In Mahadev's absence the dog's spirit protected his gear. Mahadev produced bow and arrows for the dog to function effectively. From that day on the dog became '*Sirin Shikari*'. This was the first '*shikari*'. Today tradition distinguishes 16 '*Shikaris*'. Their names and characteristics are:-

(a) **Ekle shikari** the lone hunter. He is the assistant to the others.

(b) **Lawari shikari** the colour-bearer. He leads the *shikari* troupe, carrying the seven-coloured flag white, red, blue, green, yellow, light blue and cream. *Lawa* means 'flag'. It is made of a long cherrywood staff, surmounted by an iron spearhead. Various coloured cloths are attached to an iron ring at the base of the trip. The whole thing is called barcho.

(c) **Bagali shikari** the guardian of the troupe. *Bagalinu* means 'to group together'.

(d) **Chamung shikari** this one is bad. If seen at the time of a *puja*, he may jeopardize the *puja's* success. The name is perhaps related to *chamkanu,* 'to be instant'.

(e) **Khurma Shikari** throws needles out from the earth. Inflicts illnesses such as sciatica (*khuro* means 'pin').

(f) **Namrun shikari** found under fruits and flowers. He protects. He is good.

(g) **Bajira shikari** lightning (*bajra*=lightning).

(h) **Chalena shikari** makes fun of people. He is at once good and bad. (*chale* (*adj.*) = 'to pretend')

(i) **Kheluwa shikari** plays everywhere (*khelnu*=to play).

(j) **Sarma shikari** bends people by inflicting attacks of paralysis.

(k) **Sama cherin chatmaghe shikari** if he is not properly worshipped, he inflicts fatal pains of the stomach and heart.

(l) **Raktaban shikari** causes internal haemorrhage. *(rakta*=blood)

(m) **Ban shikari** moves about like the wind and knocks people over. The paralysis is said to be caused by jungle spirit s who shoot the sufferer. (*ban* =arrow).

(m) ***Kali shikari*** black, like *bhu*t (ghost).

(o) **Solasarma shikari** if people speak badly of him, he immobilizes them on the spot.

When men are sick, the *'jhakri'* must name the 16 '*shikaris*' and designate the one responsible for sickness. A cock may be sacrificed on such occasions.

Summary

It is noticed that '*jhakri*' though does not fully embrace the same world of relationships as classical Hinduism, the '*jhakri*' and his clientele are more or less Hinduized, but are also influenced by Tibetan Mahayana Buddhism, the Mun and the Bon rites. In these accounts of cultural diffusion also, many folk tales are prevalent. The overall integration of cultural diffusive knowledge becomes definitive only when the '*jhakri*' integrally transmits his knowledge to a pupil, an exceptional event for the society as a whole. It seems clear that the pupil in turn interprets and embellishes the material in terms of his own competency and experience, and so it goes from generation to generation.

Reference

1. Subba, J.R., '*The Limbus of Eastern Himalayas : with special Reference to Sikkim*', SYMS, Gangtok, 1999.
2. Sakya, J.B., '*Buddhism and Hinduism in Nepali*' Handicraft Association of Nepal,Kathmandu, 1989.
3. Thakur, R.N., '*Himalayan Lepchas*', Archives Publishers, New Delhi, 1988.
4. Subba, J.R., *ibid.*
5. Tamang J. P. Tamang, N., '*Sikkim Biodiversity and Eco-Tourism Project*', Tadong, Sikkim, pp31, 1998.
6. Field Work in Upper Rangit Basin.
7. Field Work in Upper Rangit Basin.
8. Field Work.
9. Field Work-Ravangla Folk Market.
10. Inglis, J.T., (ed), '*Concept and Cases: Traditional Ecological Knowledge*', IDRC, 1993.
11. Govinda, L.A, '*Foundations of Tibetan Myasticism*', B.I Publications, Pvt Ltd, New Delhi, 1960.
12. Venu P. *etal,* 'Limnological Survey of Water Bodies of the Sikkim Himalaya, India', *Journal of Human Ecology* 1(2), 1990.
13. Roy, B. N., and Thapa , M. P., 'Limnology of Lake Khecheopalri', *journal of Hill Research, 9(2),* 1996.
14. —————., 'Limnology of Lake Khecheopalri-II', *Journal of Hill Research,* 1997.
15. Roy, B. N., *etal,* '*Studies on Biological Diversity of Lake Khecheopalri*', paper presented in International Conference on West Land, New Delhi and Bharatpur, 1998.
16. Raina, V.K., and Srivastava, B.S., *Tectonic Evolution of Sikkim Himalaya'*, in A.V. Sinha (ed) Geoscientific Research in Himalaya, Dehradun, 1981.
17. Bose, P.N., '*Geology and Meneral Resources of Sikkim*', the Gezetteer of Sikkim, Nature Conservation Foundation, 1989.

18. Hitchcock, J. T., and Jones, R.L., ***'Spirit possession in the Nepal Himalayas'***, Vikas Publishing House, Pvt Ltd, New Delhi, 1994.
19. Field Study in Upper Rangit Basin.
20 Burn, C.S., *The Hand Book of Folklore'*. Sidgwick & Jackson Ltd, London, 1914.
21. —————— . *Ibid.*
22. Stocks, C.D.B., *'Sikkim: Customs and Folklore'*, 1926 Cosmos Publications, Delhi, Reprint 1975.
23. Kotturan, G., *'Folktales of Sikkim*, Sterling Publishers Pvt. Ltd., New Delhi, 1989.

5

Ethnobotany of Upper Rangit Basin : Some Case Studies

SECTION-1
INTRODUCTION

Biological Ecology and Human Ecology

Ramon Margalef [1], an expert in biological ecology, defines ecology as, "the study of system at a level in which individuals or whole organisms may be considered elements of interaction, either among themselves or with a loosely organized environment matrix. Systems at this level are named ecosystems, and ecology of course is the biology of ecosystems".

Does the ecology of humans fit this definition? It does. Human individuals interact among themselves, and the interaction patterns are described as systems. People also interact with environmental matrices, which are again considered to be systems or parts or conglomeration of systems. These human ecosystems have biological properties, and according to John W. Bennet [2], they possess regularities that might form the basis of theoretical constructions. Bennet further amplifies the statement, describing man as an animal who moves on earth through its spaces and properties, consuming what he needs to survive, dealing with other biological species, largely unaware of the complicated relationship among the phenomena that he disturbs or changes - or that may disturb or change him. All this can be included in Margalef's definition quoted above.

One such related subject of biological ecology of human ecosystems is the study of ethnobotany. Human ecological research without details of ethnobotany of a given area, including the inputs of traditional plants ecological knowledge of its long time inhabitants, remains incomplete.

Paradigms of Ethnobotany and Ethnobotanical Reserch

The term ethnobotany was coined in 1895 by the University of Pennsylvania by botanist John Harshberger (1896) of the University of Pennsylvania. E Wade Davis[3] elaborates that ethnobotany as a practice goes back in Western and Eastern traditions to prehistoric times.

The Scope of a field of inquiry is reflected in the questions essential for ethnobotanical research. Specific questions asked by ethnobotanists include the following (Lanis B. Alcorn[4], 1995) :-

(a) What plants are available ?

(b) Why are they available ?

(c) What plants are recognized as resources ?

(d) What social, political, biological, economic and ecological factors cause particular plants to be perceived as resources ?

(e) How does the use of certain set of resources influence the use/ availability of others ?

(e) How is ethnobotanical knowledge distributed among the human population ?

(g) What do people think about plants ?

(h) How do they differentiate and classify elements of their natural environment ?

(i) From what resource zones are plant products harvested ?

(j) How are they used ?

(k) What are the economic and financial benefits derived from plants?

(l) How are the populations of resources maintained ?

(m) What effect does their management have upon the structure of local vegetation ?

(n) What effects does their management have upon the structure and functioning of local institutions ?

(o) What factors influence resources management decisions, and thereby affect local plant populations ?

(p) How have human activities and their consequences influenced the evolution of local plant populations ?

(q) For what purpose are resources needed ?

(r) To what stressors are human populations adapting ?

(s) Are human choices of particular resources adaptive ?

(t) If, so how are they adaptive ?

(u) How are human adaptive strategies affected by change ?

(v) What changes are presently occurring and what changes have occurred in the past ?

As can be seen from these questions, modern ethnobotany is

concerned with the "totality of the place of plants in a culture" (Ford[5], 1878). It is the study of plant human inter-relationships embedded in dynamic ecosystems of natural and social components. The above questions formed the basis during the field work proceedings of present research work for this thesis.

The object of ethnobotanical inquiry is actually a sort of "Text" (Ricoeur[6], 1971), the meaning of which is derived partially from the natural, social and cultural contexts in which that text is played out. An ethnobotanical text evolves around human community's use and management of vegetation. Ethnobotanical resources and their non timber uses in areas of Yuksom have been studied and listed in this chapter.

Ethnobotany of Traditional Agroecosystems and Conservation of Wild Species

The land managed under a traditional system of agriculture usually contains many different patches gardens or fields of various ages containing different combinations of cultivated and wild plants according to the farmer's decisions and the needs of his or her family. The wild vegetation diversity created by traditional systems apparently functions to protect the agroecosystem from pest outbreaks (Speight 1983 ; Alteieri [9]1984. Indigenous maintenance of wild resources is , in effect , indigenous "conservation" of wild species. Traditional farming systems often create fallows and other corridors or patches of natural vegetation that act as resource zones, biodiversity conservation units and gene banks for the regeneration of natural communities (Lynch and Alcorn[10] 1991). The study of resource conservation is an area of ethnobotanical inquiry that reaches into the heart of ethnobotanical text and contributes to reaching the applied goals of the ethnobotanical endeavor. By identifying ecologically adaptive aspects of resource management, ethnobotanists contribute to the development of resource- conservative, sustained-yield agroecosystems (Hart 1980, Wilson and Kang[11]1981).

The Human Ecology Connection

Ethnobotany plays a major role in facilitating the study of human ecology. Modern ethnobotanical data are often used as analogues for interpreting past plant-human inter relationship. (Dimbley[12] 1978). By providing data about the plants of the ecosystem, ethnobotanists do contribute to more sophisticated analyses of human ecological relations.

(Boster[13] 1984). As applied human ecologists, economic and social anthropologists have been granted an expanded role in rural development projects (Barlett[14] 1980; International Potato Centre[15] 1984). Their study and research into ethnobotanical aspects becomes that much amplified.

Ethnobotany of Medicinal Plants

One of the subdivisions of modern ethnobotany in the study of applied human ecology is the study of aboriginal knowledge of the physical properties of plants and the accompanying familiarity with those components of the vegetal environment that permit indigenous peoples to use various plants for their presumed medicinal value. This subdivision is now known as ethnopharmacology (Schultes and Reis[16] 1995).

In the dawn of human kind, the art of curing was essentially magical (Narenjo[17] 1984). Few plants were used and those employed were usually psychoactive ones, now known as magic or psychedelic plants. Later, empirical medicine arose, using many plants for the treatment of various affiliations. Scientific medicine arose much later. The 17th century, in particular opened the doors to true scientific understanding of medicine. Although the art of curing was refined, plants continued to occupy pre-eminent position. Until 1930, around 90 per cent of the official medicines were of plant origin (Swain[18] 1972). As previously analysed by many authors (Fransworth and Bingel 1977; Von Reis Altschul 1973 ; Balendran 1985; Duke 1985)[19-22], there are many powerful reasons for doing broad ethnobotanical, ethnopharmacological and also clinical therapeutic research. Among these reasons the following could be cited(Klayman[23] 1985):-

(a) To rescue knowledge in imminent danger of being lost.
(b) The utility of plants in current therapy.
(c) To find new molecular models in plants.
(d) The usefulness of plants in the development of physiopathology.
(e) The wide use of plants in folk medicine.
(f) To obtain intermediate chemicals.

A case study of medicinal plants, including ethnic group wise case study, and plants in Yuksom and Dzongri areas is included in this chapter.

Applied Ethnobotany in Home Gardens

Home gardens (also known as compound farms, homestead and

mixed gardens) are usually the small plots of land surrounding the house. They are found in traditional communities all over the world, and are one of the more intensively cultivated parts of an overall farm. Home gardens are important agroecosystems and are a source of subsistence and cash resources. They are characterized by a mixture of annual or perennial species grown in association. (Godbole[24] 1998).

Home gardens commonly exhibit a layered vertical structure of trees, shrubs and ground cover plants which recreate some of the features of nutrient recycling, soil protections and effective use of space below and above the soil surface. They also act as a repository and testing site for uncommon species and varieties of plants(Padoch and Jong[25] 1991) and can be used to spread framework output and income more evenly throughout the year (Ninej[26] 1984).

This chapter includes a study on Home Gardens as available near the confluence of Rathong Chu and Rangit Chu in the areas of Tashiding and Western slopes of Ravangla.

Local Folk Markets for Sale of Wild Plants/Plant Products

Rural folks, especially in mountain regions are largely dependent on wild plants/plant products for their existence. The local markets are full of wild vegetables, fruits and medicinal plants. They make a variety of fish traps, hats, agricultural implements, containers and other household articles and sell them in these markets.

These markets are of permanent nature in large towns, while in smaller places these are held on a fixed day. Interestingly most of these markets are run by women. The buyers are often from neighbouring localities (Saklani and Jain[27] 1994). A case study of Wednesday market traditionally held at Ravangla forms a part of this chapter.

Eating Habits: Traditional Fermented and Non Fermented Foods

In the Sikkim Himalayan regions including the Upper Rangit Basin there exists a unique dietary culture of different ethnic groups of people. They use traditional fermented and non fermented foods and beverages to provide basic components of diet with diverse characteristics of nutrition, flavour, palatability and texture.

A survey conducted randomly in the Upper Rangit Valley representing the ethnic communities of the Nepalis, the Bhutia and the Lepcha was carried out. The same has been listed in this chapter.

Data Presentation of Sikkim Ethnobotanical Resources

It is interesting to note that various parts of plants, such as seeds, flowers, bark, fruits, rhizomes and tubers are used for different purposes. In some cases whole plants are also used. An inventory of such plants, and their uses according to different parts of the plants for the state of Sikkim is presented at the end of the chapter. The research for this inventory has been carried out from available literature, and through field work done in collaboration with local government and non government organisations.

SECTION II
HOME GARDENS IN UPPER RANGIT BASIN

Introduction

Home gardens can be defined as the land surrounding a house on which a mixture of annual and perennial plants are grown, together with or without animals, and largely managed by the household members for their own use or commercial purposes. Brownrigg[28] (1985) defines the term as, "A supplementary food production system by and for the members of a group of people with rights to the land, who eat meals together regularly. "Fernandes and Nair[29] (1986) state that the term Home Garden can mean anything from growing vegetables behind house to complex multistoried systems. They defined the term as "Land use practices involving deliberate management of multipurpose trees and shrubs in intimate association with annual and perennial agriculture crops and, invariably livestock's, within the compound of individual houses, the whole crop-tree-animal unit being intensively managed by family labour".

Home Gardens are an ancient and widespread agroforestry system in Eastern Himalayas including Sikkim. The cropping pattern changes with altitude. The size and shape of the gardens is dictated by gradient, contours and terrace availability. The prevalent climatic condition ranges from subtropical to alpine with major area under temperate climate in the Upper Rangit Basin of Sikkim. One single spur rising Westwards from Rathong chu upto the watershed on Singalila Ridge experiences climatic conditions from tropical to subtropical to temperate and finally to alpine at the summits of the ridge. As such ecological situations vary greatly at very short distances which limits the adaptation of crop species and their varieties within a narrow range.

This has the advantage of growing many kinds of fruits, seasonal and off seasonal vegetables and other commercial crops in the hills at different altitude and in different seasons (Subba[30] 1984).

Home Garden Structure

The vertical stratification of vegetation within Home Gardens has long been recognized as a characteristic feature, although the variation of height within any one stratum has led to some arguments as to the distinctness of the various strata recognized by various authors. Barrau[31] (1961), Michon [32] (1983), Altieri and Farrell [33] (1984), Fernandes et al[34] (1984), Okafor and Fernandes[35] (1987), Oduol and Aluma[36] (1990), Millat-e-Mustafa[37] (1996) give schematic presentations of vertical structures from various geographical regions and observe that the canopies of most Home Gardens consist of strata between two and six.

General summary of strata as found in the Home Gardens in Tashiding and in the areas immediately adjacent on the East and West of Rathong Chu and Rangit Chu just North of their confluence is as under :-

< 1m	Vegetables, tubers and roots including cardamom broom grass and flowers.
1-3 m	Food plants e.g., lemon, banana, Papaya and broom grass.
3-5 m	Oranges and Saplings of fruit/timber trees all growing taller.
5-7 m	Bamboo, timber/fruit trees, some growing taller.
7-9 m	Mango trees, timber trees, fodder trees and bamboo.
>9 m	Timber trees (Oak and Chestunt) and bamboo.

It is stressed that the above strata are dynamic and there is a constant recruitment from one stratum to another.

Floristic Diversity in Home Gardens of Study Area

Diversity is defined as, "Many different species and their intensity of interactions occurring in a small space at one time". (Harmer[38] 1991) Indeed this definition neatly encapsulates the concept of diversity in Home Gardens, since there is a great variety of interactions taking place vertically, horizontally and temporally within one garden often of less than one hectare.

Vegetables Diversity. The vegetable cultivation extends from river beds at the confluence to elevations upto 2000 m above sea level at Tashiding and on either side of the confluence. Following vegetables are grown :-

(a) **Kharif Vegetables.** Brinjal, chilli, capsicum, franch beans, cow pea, bhindi, cucumber, pumpkin, bottle gourd, sponge gourd, ridge gourd, snake gourd, bitter gourd, sweet gourd etc.

(b) **Rabi Vegetables.** Cabbage, cauliflower, broccoli, garden pea, french beans, radish, carrot, turnip, palak, amaranthus, methi, leafy mustard, lettuce, celery, coriander, leafy onion, garlic, leak, potato, colocasia, elephant's yam, etc.

(c) **Off-season Vegetables.** Cabbage, cauliflower, broccoli, brussel sprouts, radish, carrot, tomato, green pea, franch beans, palak, leafy mustard, chayote, cucumber, amaranthus, leafy onion, summer potato, capsicum, asparagus, bamboo shoots, fern shoots, stinging neetles, mushroom etc.

Fruits Diversity. Mango, banana, peach, plum, orange, strawberry, jackfruit, papaya etc. Apples are grown in the upper slopes of Yuksom.

Feed and Fodder Diversity in Home Gardens. The farmers mostly feed animals with fodder tree leaves and fodder grasses, in addition to agricultural crop residues and fodder crop concentrates. Tarnutzer and Batting[39] (1994) found that 87 per cent of the farmers were growing grasses on bunds in agricultural fields, steep slopes etc.

Fodder Grasses. Bush type fodder grasses, average per farm and per centage of farmers involved are given in Table 5.1 below :-

Table 5.1
Bush Type Fodder Grasses

Grass Species	*Bushes per farmer (Average)*	*Farmers Involved (%)*
Amlisho	527	62
Napier	33	31
Guatamala	2	9
Bamboo	74	17
Narkat	3	1

Paljor[40] (1998)

Fodder Trees. All the farmers on the spurs extending Westwards and Southwestwards from Ravangla and Ralong respectively are having fodder trees with a range of 72-99 per Home Garden. The different types of fodder species grown in and around Home Gardens are presented in Table 5.2 below :-

Table 5.2
Fodder Trees in Home Gardens and Farms

Species	*Local Name*	*Species*	*Local Name*
MAJOR FODDER TREES		**MINOR FODDER TREES**	
Artocarpus lakoocha	Barar	Brassaiopsis mitis	Chuletro
Ficus roxburghi	Nevaro	Bauhinia purpurea	Taki
Ficus cunia	Raj-khenew	Castanopsis tribuloides	Katus
Ficus calarata	Lutey-khenew	Celtis tetrandra	Khari
Ficus nemoralis	Dudilo	Evodia fraxinifolia	Khanakola
Ficus Saemoarpa	Bar	Ficus hookeri	Chbra
Saussuria nepaulensis	Gogun	Litsaea polyantha	Kutmero

Factors Affecting Diversity

A variety of factors affecting diversity in the study area came to light. The factors are corroborated by the literature available from earlier research by eminent ethnobotranists. Mergan[41] (1987) cites personal choice of the farmers, but more often external forces come into play. Soemarwoto[42] (1987) has listed a concise catalogue of factors :-

(a) Better financial position leading to fewer food crops and more ornamentals.
(b) Scarcity of labour prompting farmers to grow more labour saving perennials.
(c) Proximity of markets, influencing farmers to include cash crops to their gardens.
(d) Population pressures, which decreases the size of land holdings, subsequently although cropping intensity increases.
(e) Prices of commodities usually cause the sacrifice of species diversity.

The management skills of individual farmers acquired empirically over generations, in dealing with the diversity are constantly emphasised. Most farmers possess traditional ecological knowledge, such that they are often able to choose the correct 'niche' for each plant' depending on the gradient of light and altitude. This seems to correspond to the ecological niche in the natural forest.

There are wide range of uses from gardens such as :-

(a) Multipurpose tree crops provide shade for cardamom plantations and for sitting under.
(b) Parameter plantations act as living fences, fodder and mulch, bee forage, fuel wood, fruit, timber and poles.
(c) Other components provide for home consumption and also for sale.

(d) Certain varieties are used for arts and crafts, fibres for ropes, mats and even ornamentation.

(e) Diversity of plants reduces soil erosion. There are two categories of vegetation in relation to soil erosion. These act as barriers against erosion by rain. The ground litter acts as surface cover against splash and surface erosion due to winds.

(f) Home gardens with biotic diversity are a valuable genetic resource. They are a valuable gene pool for breeding and improvement programmes since selection processes, both natural and human, have occurred over the years and generations.

(g) Diversity acts as a safeguard against pests and disease. (FAO[43] 1989).

Animal Diversity

In addition to the variety of nutrition derived from food crops, diets are further augmented by animal products. Brownrigg's literature review[44] (1985) indicated that all types of Home Gardens support domesticated animals.

Same is true for the Upper Rangit Valley study area. Poultry, pigs, cows, goats, sheep and buffalo are supported by most gardens in the area. It was however felt during the field visits, that far less attention is paid to animals than to plant species.

Labour Requirements for Home Garden Management

Larger Home Gardens of the size upto one hectare demand half hour to two hours of daily labour in the study area. The smaller gardens demand upto one hour of daily labour. 35-45 days of family labour per year are utilized at the initial establishment of Home Gardens. Subsequently approximately 30 days of the year are utilized towards tending of the gardens.

There seems to be a clear sharing of tasks between women and men for the management of the Home Gardens. The initial heavy labour is done by men. Later women spend about 10 per cent of their working time on the gardens, while men work for 4-5 per cent of their working hours. Rich farmers from erstwhile Kazi families employ hired labour.

Interviews and Questionnaire Methodology

There were guided interviews conducted for checking different management issues for which farmer's opinion was wanted. The interviews were guided to cover the key topics on the checklist, while leaving rooms to pursue relevant subjects brought up by the farmers themselves. The semi-structured questionnaire is mentioned below :-

Question 1. "What planting materials do you use for your Home Garden Plants"?

This was to elicit farmer's knowledge about regeneration procedures with different Home Garden plants. The relative advantages and disadvantages of various type of planting material for different species were ascertained through further questions.

Question 2. "What are the sources of different planting materials?"

This was to establish the relative contribution of different sources of planting materials.

Question 3. "Do you follow any criteria to select mother trees to collect planting materials ?"

This retrieved farmer's knowledge about the introduction of improved varieties of species in the Home Gardens. If the farmer said 'yes', then details of the criteria for mother tree selection and the types of species for which mother trees are selected were sought with more questions.

Question 4. "What sizes of seedling are available to plant and which one do you prefer and why ?"

This was to explore farmers silvicultural knowledge about different sizes of seedlings.

Question 5. "Do you adopt any spacing at the time of planting ?"

This question was to gather information on horizontal arrangements of species in the Home Gardens. When the answer was 'yes' , further questions were asked to ascertain the planting, spacing and the reasons for the same.

Question 6. "Do you carryout weeding, lopping, pruning, thinning, coppicing, pollarding in your Home Garden ?"

Weeding and thinning determine the horizontal structure of the Home Gardens, while pruning determines how the farmers regulate sunlight in the Home Gardens. Division of labour between men and women in these contexts was also discussed and answers sought.

Question 7. "Do you water and manure your Home Garden plants?"

This question was posed to produce an idea of the effort given to managing the Home Garden. When the reply was 'yes' the frequency, quantity and calendar of watering and manuring, and the name of the species to which these were applied was determined with further questioning.

Socioeconomic and Ecological Aspects of Home Gardens

Socioeconomic aspects and community dynamics were closely observed.

These were found to be important factors for maintaining the traditional systems. The interaction dynamics were mainly qualitative. The social, cultural, economic and management aspects are given at Figure 5.1.

Figure 5.1 Socio Economic Aspects of Home Gardens

Social	Economic	Cultural	Management
Social Functions	Productivity	Attitudinal Change	Specific Practices
Role of Women	Utilisation	Cultural ethos	Calendar of activities
Social Relationships	Provisions for adverse conditions	Social trends	Decision-Making
		Social functions	Working Pattern

The aspects of human ecology of the Home Gardens form an array and an internet of inter-relationships. Figure 5.2 overleaf amplifies the interactions.

Figure 5.2 Ecological Aspects of Home Gardens

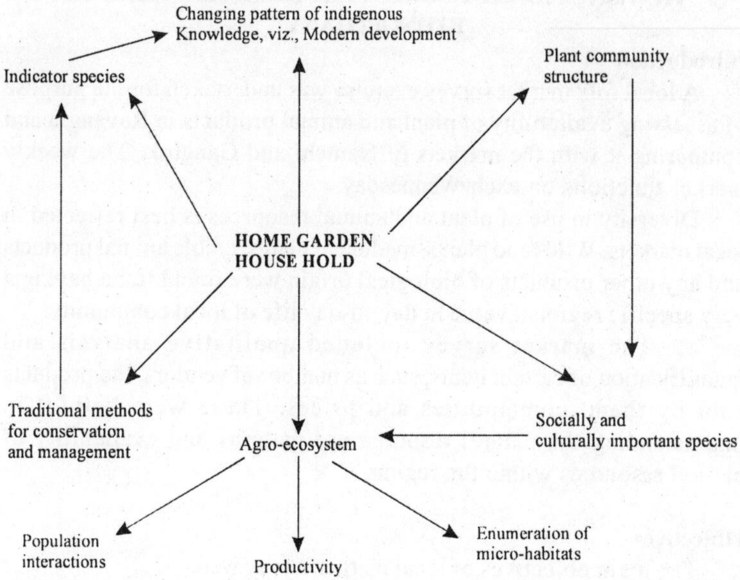

Conclusion

Historically the Home Gardens in Sikkim have been little appreciated, but much used. Sikkim Home Gardens are traditional resource management technologies that have recently been hailed as highly productive and largely sustainable agroecosystems.

The Home Garden is but one of a very large array of agroecosystems that traditional communities maintain.

Many of the species that are grown in Home Gardens are also found in agriculture fields and on the fallow lands adjacent to the households. Many are collected from nearby forestlands. Thus Home Gardens of the Sikkim indigenous communities deserve to be examined in far greater detail. The study of Home Gardens could be used as a tool to develop methodologies for the application of traditional knowledge in conservation and community development.

SECTION - III

LOCAL FOLK MARKET: A SURVEY OF MODES OF TRANSACTIONS OF FOREST PRODUCE RELATING TO ETHNOBOTANY

Introduction

A local folk market survey exercise was undertaken for the purpose of assessing availability of plant and animal products in Ravangla and comparing it with the markets of Namchi and Gangtok. The weekly market functions on each Wednesday.

Diversity in use of plant and animal resources is best reflected in local markets. Wild food plants, medicinal plants, edible animal products and any other products of biological origin were found to be having a very specific regional value in day-to-day life of local community.

The market survey included qualitative analysis and quantification of certain items, such as number of vendors, the products sold by them, commodities and prices. These were helpful in understanding the natural resource use patterns and availability of natural resources within the region.

Objectives

The main objectives of local market survey were :–

(a) To assess the local use of resources.

(b) To determine natural resource use patterns for specific commodities.

(c) To understand the dependence of Ravangla locals and population of nearby rural folks on the surrounding resource areas.

(d) To evaluate the sustainability of the natural resources the factors responsible for it.

Methodology

The simple methodology used for this local market survey mainly involved participatory observations of the vendors and buyers along with use of semi structured and structured interviews. These were composed with the surveys of the Home Gardens to check the origin of various commodities in the market.

Observations

Commodities Diversity

(a) A variety of plants, animals and other products of biological origin were being sold.

See table 5.3 and 5.4 for some important wild plants available at Ravangla and their comparison of price range with the market price at Gangtok and Namchi. The prices at Ravangla are minimum, perhaps due to minimal labour costs in carting the products. Namchi is cheaper than Gangtok for the same reasons. The market at Gyalshing is quoted to be the cheapest in this regard.

(b) These products were either harvested from the Home Gardens or gathered from the wild.

(c) There was a clear division of items available from the Sikkim hill/mountain regions and from West Bengal plains.

(d) Most commodities were being sold raw ie fruits, vegetables, meat and fish. Dehydrated fish were available in plenty. Fermented bamboo shoots and Rhus seeds were however sold processed.

Vendors

(a) Permanent vendors occupy the main shelters erected in the market.

(b) Temporary vendors are allowed to occupy specific space allotted by the Sub Division office at Ravangla.

(c) The number of temporary vendors far exceeds the permanent vendors.

(d) The permanent vendors are generally Bhutias, while most temporary vendors are Nepalese of different castes, and Tibetans from the Tibetan Refugee Settlement named Kunphenling on the Southern outskirts of Ravangla. Marwaris who occupy permanent shops in main market do not open stalls in the weekly Wednesday market. Bengali and Bihari tradesmen who frequent other weekly markets in Sikkim also display utility items such as readymade clothing items, shoes and other household goods brought from wholesale market in Siliguri. Imported items (Some of them locally made imitations) are also on display in some stalls.

Table 5.3
Some important wild plants available in the weekly market at Ravangla in South Sikkim

Plant species	*Local Name*	*Time of year*	*No. of retailers*	*Ravangla Market Price (Rs)*
Bassia butyracea	Chiuri	May	5	3 per kg
Agaricus Sp	Chayo	Jun-Sep	2	15-25 per kg
Castanopsis tribuloides	Katus	Oct-Dec	2	30 per kg
Cinnamonum Sp	Sinkauli	Whole Year	4	30 per kg
Dendrocalamus hamiltonii	Tambu	Jun-Oct	2	12 per kg
Dioscorea Sp	Bantarul	Jan-Mar	5	5 per kg
Diplanzium Sp	Ningro	Mar-Aug	20	5 per kg
Elaeocarpus Sikkinmensis	Bhadrase	Apr-Jun	1	10-15 per kg
Elaeagnus Latifolia	Malido	Feb-Mar	2	8 per kg
Emblica officinalis	Amala	Oct-Jan	3	5 per kg
Ficus benjamina	Kabra	Feb-Mar	2	8-10 per kg
Heracleum Wallichii	Chimfing	Aug-Nov	3	8 per unit
Junglans regia	Okhar	Sep-Nov	4	20 per kg
Machilus edulis	Pumsi	Jan-Feb	1	5 per unit
Rhus Succedanea	Timbur	Aug-Sep	2	3 per 50 gms
Rumex nepalensis	Halhale	Whole Year	2	10 per kg
Spondias axiliaries	Labsi	Oct	2	5 per kg
Swerlia chirata	Chirauto	Jun-Dec	2	15-20 per kg
Tamarindus indica	Titri	Mar-May	3	4 per kg
Terminalia belerica	Barra	Whole Year	2	15-20 per kg
T Chebula	Harra	Feb-Mar	7	20 per kg
Urtica dioica	Sisnoo	Dec-Mar	7	2 per unit
Viscum album	Harchur	Whole Year	4	5-10 per kg

Table 5.4
Some important wild plants available in the weekly market at Namchi and Gangtok in Sikkim

Plant species	*Local Name*	*Time availability*	*Namchi*		*Gangtok*	
			No of Retailers	*Market price (Rs)*	*No of Retailers*	*Market price (Rs)*
Agaricus sp	Chayo	Jun-Sep	6	10-20*	8	15-25*
Bassia butyracea	Chiuri	May	6	3-5*	5	5-8*
Castanopsis tribuloides	Katus	Nov-Dec	7	40*	4	40-60*
Cinnamonum sp	Sinkauli	Whole year	10	30*	11	40*
Dendrocalamus hamiltonii	Tambu	Jun-Oct	4	12*	6	15*
Dioscorea Sp	Bantarul	Jan-Mar	5	12*	6	10*
Diplanzium Sp	Ningro	Mar-Aug	10	10-15*	13	15-20*
Elaeocarpus Sikkimensis	Bhadrase	Apr-Jun	3	10-20*	3	15-20*
Elaeagnus Latifolia	Malido	Feb-Mar	-	-	6	10*
Emblica officinalis	Amala	Oct-Jan	3	4*	-	-
Ficus benjamina	Kabra	Feb-Mar	5	8-10	2	15-20*
Heracleum Wallichii	Chimfing	Aug-Nov	4	10+	3	80+
Junglans regia	Okhar	Sep-Nov	8	20=	14	50=
Machilus edulis	Pumsi	Jan-Feb	8	15*	12	20-25*
Rhus Succedanea	Timbur	Aug-Sep	3	2-3+ +	5	3-5 + +
Rumex nepalensis	Halhale	Whole year	4	10*	6	2 + +
Spondias axillaries	Labsi	Oct	7	4*	10	5*
Swerlia chirata	Chirauto	Jun-Dec	4	15-20*	3	20-30*
Tamarindus indica	Titri	Mar-May	8	5*	10	8*
Terminalia belerica	Barra	Whole year	3	20*	2	-
T Chebula	Harra	Feb-Mar	5	20=	5	25-30=
Urtica dioica	Sisnoo	Dec-Mar	7	5 + +	6	3 + +
Viscum album	Harchur	Whole year	6	5-10*	4	15*

* per kg, = per 100 units + per 50g + + per 100 g

(Adopted from G.B Pant Institute[46] Annual Report 1996-97).

(e) There is no permanent shelter available. The market functions in the main bazaar square. The square functions as volleyball and basketball playfield on non-market days.

The Indian Republic Day and Independence Day functions are also celebrated in the market square. This is also the focal point of all election rallies.

Buyers

(a) Buyers include persons from all walks of life and from all ethnic groups. Government servants including soldiers from local military camp also frequent the market.

(b) The local villagers demand priority is house hold goods, clothing and other items brought from the plains of West Belgal.

(c) Ravangla locals mainly buy vegetables, meat, poultry and other ethnic commodities on display.

(d) Some individuals of both sexes frequenting the market freely indulge in consumption of 'chhang' and 'rakshi' (a locally processed fermented alcoholic drink) and other Sikkim manufactured foreign liquor such as beer, rum, whisky and brandy. Beer is the most preferred drink with the local ethnics and the Tibetans of Kunphenling Settlement. Violence due to alcohol consumption is conspicuously absent.

Role of Women in the Local Markets.

(a) It was observed that 80-85 per cent of vendors were women.

(b) Women vendors reported that men were engaged in collection of commodities from the wild.

(c) Women looked after management, cultivation, harvesting and processing of products.

(d) Women also reported that as a result of trading, they do achieve a certain control over the household economy, but it is usually restricted to using the money gained from marketing for buying necessary commodities such as clothes toiletries, bamboo baskets etc.

Rhythms of the Market. It is very interesting to note the rhythms of the market, ie; the changes that occur at regular intervals:- (Also see photofeature of market on next page)

(a) During the first two hours there is a rush for sale of items such as perishable vegetables and freshly slaughtered meat. Marketing activities are at a peak during these two hours.

(b) After this initial sale range of commodities such as dried fish, spices, poultry meat, fermented bamboo shoots and *chhang* etc become available.

(c) The household goods stalls have a more or less regular sale during the entire day.

(d) The vendors in the market begin to thin out by 3 PM during winter months, and by 4 PM during summer months. By 5 PM, barring hangers on, the market is virtually empty.

Statutory Framework

The first attempt for managing the bazaar activities under a statutory framework was in the form of Sikkim Bazaar Committees Act,

Rhythms of Ravangla Market

Most Vendor's are Women

1969 and Gangtok Municipal Corporation Act, 1975. Both of these Acts were repealed by the Sikkim (Repeal and Miscellaneous Provision) Act, 1985. Under this Act, the functions, powers and duties of the Bazaar Committee and Gangtok Municipal Corporation were transferred to and vested in the State Government in the Housing and Urban Development Department (Mahendra Lama[45] 1994).

Strategy for Development

The role of the informal sectors has to be recognized as a potential for employment generation. A state like Sikkim which has an agrarian base and ample natural resources for development, would have to encourage the local market system for its products and employment generation, mainly for the following reasons :-

(a) The markets act as a first contact point for local residents, petty traders and small farmers.

(b) They provide place for distribution of local produce and exchanging rural surplus.

(c) They provide an opportunity to petty traders and small farmers for earning their livelihood.

(d) They provide place for buying daily necessities as well as farm supplied foods and implements, at less expensive prices, especially to the small farmers.

(e) They play a vital role in well being of small farmers, petty and mobile traders.

(f) They are the best mechanism of serving a dispersed population, mostly agrarian in nature with limited demand.

(g) The means of transportation still being difficult, and many commodities being of perishable nature, dispersed market places, not involving big investment for development, are the best alternatives.

Conclusion

Weekly market in semi-urban townships such as Ravangla and other areas of Sikkim focus on local demand. Such local market surveys could be used to identify local resource needs. Such studies would also help to assess the role of existing agroecosystems in natural resource use and management.

Quantitative studies carried out during such market surveys will help to evaluate the sustainability of resource utilization, and may also be used to develop enterprise potential for economic and environment sustainability. Local markets and transactions within them throw light on cultural aspects of nature resource use patterns.

Suggestions/Recommendations

A number of commodities such as wild vegetables, fruit and animals have a high demand and there is much scope for domestication of such items. This will afford following advantages :-

(a) Domestication will help to reduce the pressure on existing forest and ensure a continuous supply.

(b) Some vegetables and fruits have a high demand but cannot be marketed because of non-availability of proper post harvest technology and storage facilities. Specific research is needed to develop post harvest technology and storage facilities.

(c) The market is unorganized. Domestication, improved post-harvest techniques and organized markets will help to achieve proper and sustainable resource use.

The statutory laws could also be simplified for the betterment of the simple village folk. The banking and loan procedures should be made simple, and the by laws be made clearly understood to the rural vendors.

SECTION-IV
DIETARY CULTURE OF LOCAL ETHNIC COMMUNITIES OF THE STUDY AREA

Introduction

Ecological factors to a large extent determine which foods are available in a culture. These factors include climatic conditions, vegetal zones, local faunal and floral diversity, agroecosytems, socio-religious factors and the application of traditional ecological knowledge.

The term 'culture' includes symbolic, social and economic factors, which in interaction with sensory tastes and preferences shape food patterns and influence the selection of foods. The collective eating habits of a cultural group forms a 'cultural cuisine'. The term cultural cuisine is used to describe, "The culturally elaborated and transmitted body of food related practices of any given culture" This includes (Messer[47] 1989):-

(a) The selection of a set of basic (staple or secondary) foods.

(b) The frequent use of characteristic set of flavourings.

(c) The characteristic processing (eg chopping and cooking) of such foods.

(d) The adoption of a variety of rules dealing with acceptable foods and combinations, the social context of eating and the symbolic use of foods.

Within cuisines, people select foods to conform particular styles of eating, as well as cultural concepts of what items are edible and preferred, and how they should be prepared and served. People eat discontinuously; that is, food is structured into breakfast, lunch, dinner and supper in a 24 hour cycle. The cycle differs for each culture or ethnic group, and also for gender and age groups.

Dietary Structure in Sikkim

In the Sikkim Himalayan regions, including those of the Upper Rangit Basin a unique dietary culture exists. The inhabitants of the Upper Himalayan belt (above 2500m) have common dietary habits as the Tibetans. Balance of the inhabitants, which form over 95 per cent of the total population use traditional fermented and non-fermented foods and beverages to provide basic components of diet with diverse characteristics of nutrition, flavour, palatability and texture. For many centuries, the people of this region prepare and consume varieties of traditional fermented and non-fermented food products (Tamang et al 1988; Tamang 1988) [48-49].

The *Bhat-dal-tarkari-achar* (rice-legume soup-curry-pickle) constitute the daily food supply in the meal. Rice is the staple food. The cereal-based fermented food products are mostly consumed as alcoholic beverages.

FERMENTED FOODS

General

Fermented foods are defined as the products that have been subjected to action of selected micro-organisms in the substrates changing texture, aroma and flavour, and are acceptable for human consumption (Kwon 1994, Steinkraus 1994, Campbell-Platt 1994) [50-52]. The people of the Sikkim Himalaya developed traditional food processing technology for preparing fermented foods from locally available substrates largely governed by the ethnic preference, the agro climatic conditions, socio-cultural ethos and religion. People might not be knowing what was going on during fermentation in terms of scientific knowledge, but they certainly knew what they had to do to get desired products . This is how indigenous fermented foods emerged as the basic components of diet and reflect a dietary-culture of the various ethnical groups of people in the Sikkim Himalaya (Tamang [53] 1995b). More than seventeen varieties of indigenous fermented foods are traditionally prepared and consumed by the people at households level for long centuries in Sikkim (Table 5.5). These inexpensive culturally acceptable traditional foods supplement enhanced nutritional qualities, palatability, imparting appealing flavour and increased shelf-life. Some of the common fermented foods have been cited here. Figure 5.3 depicts distillation apparatus, and drinking vessels for consumption of '*raksi*'.

Table 5.5
Indigenous Fermented Foods of thte Sikkim Himalaya

Fermented food	*Substrate*	*Nature of use*
Common fermented foods		
Kinema	Soybean	Cooked soybeans showing stickyness with Typical flavour; side dish
Gundruk	Leafy vegetable	Dried, sour-acidic taste; soup, pickle
Sinki	Radish root	-do-
Selroti	Rice/wheat flour	Pretzel-like, deep fried; confectionary bread (doughnut)
Dahi	Milk	Thick-gel; savory (yoghurt)
Mohi	Milk	Butter-milk; non alcoholic drink
Chhurpi	Milk	Soft-mass is used for curry, pickle Hard mass is consumed as masticatory
Sukako masu	Mutton/beef/pork/ Yak	Smoked meat; curry
Common fermented beverages		
Marchaa	Rice, wild herbs, Spices	Solid, dry, while cake-like starter culture to ferment beverages
Kodo jaanr	Finger millet	Midly alcoholic, slightly acidic-sweet Beverage.
Bhaate jaanr	Rice	-do-
Makai jaanr	Maize	-do-
Raksi	Starchy substrate	Distilled alcoholic drink
Lesser-known fermented foods		
Masauyra	Black gram	Ball-like hollow; spicy condiment
Mesu	Bamboo shoot	Sour-acidic flavour; pickle.
Khalpi	Cucumber	-do-
Somar	Milk	Soft-mass paste, brownish colour with Strong flavour; condiment

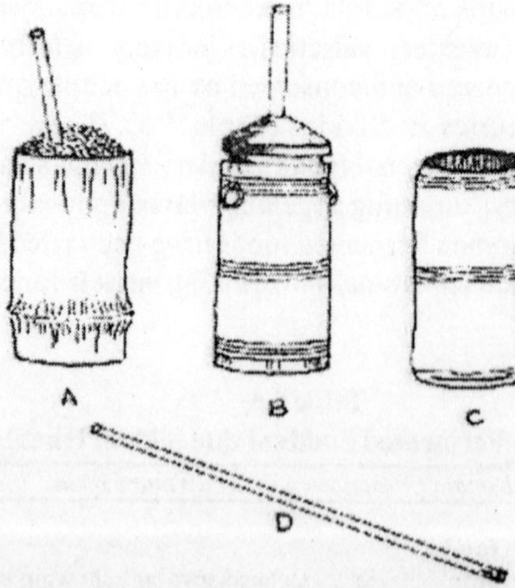

Toongbaa made up of (A) bamboo vessel (B) wood in decoration, (c) mud, (d) piping to sip extract of kodo ko jaanr from Loongbaa

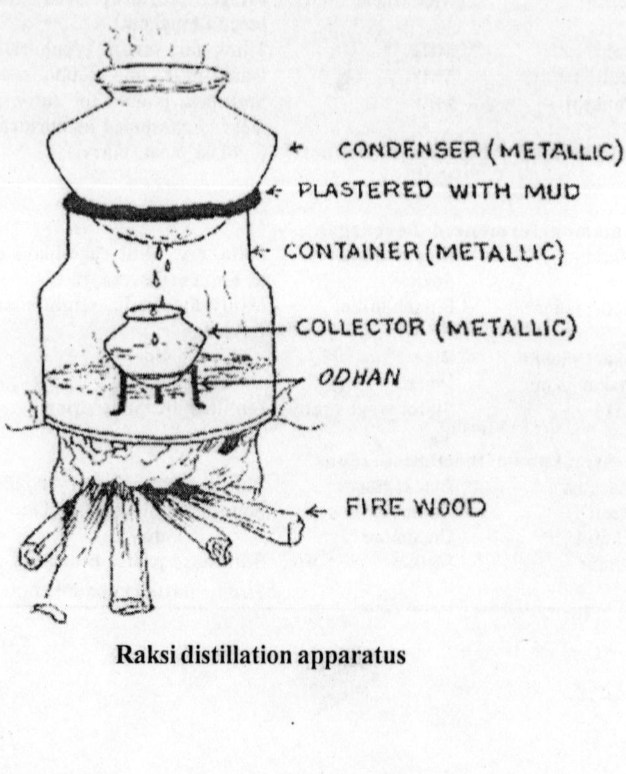

Raksi distillation apparatus

Kinema

Kinema is an indigenous fermented soybean food which serves as a sustainable supply of low-cost source of high protein food in the local diet. Kinema curry is delicious local cuisine eaten along with boiled rice in main meals(Tamang [54] 1996).

Synonym of kinema in local language

Kinemba (Limbu)
Hokuma (Rai)
Bari (Bhutia)
Satlyangser (Lepcha)

Similar products like kinema

Hawaijar (Manipur in North-East India)
Bekang-um (Mizoram in North-East India)
Troombai (Meghalaya)
Akhoni (Nagaland)
Itohiko-natto (Japan)
Tooa-nao(Northern Thailand)
Chungkook-jang(Korea) (Tamang [55] 1998)

Gundruk and Sinki

Gundruk and Sinki are very popular fermented vegetable products traditionally prepared during winter when perishable vegetables are plenty. Gundruk and Sinki are consumed as soup and pickle in main meals. The quality attributes to gundruk and sinki basically depends upon the typical flavour and acidic taste.

Synonyms of gundruk

Gunduru (Newar)
Mashit (Rai)
Kandruk (Limbu)

Synonyms of Sinki

Sinke (Newar)
Samchetwa (Rai)
Thektubalawak (Limbu)

Socio-economy of gundruk and sinki

Some rural people are economically dependent upon these products. They sell gundruk and sinki usually during winter in the local markets. Cost of gundruk is Rs 60/- to Rs 70/- per kg, and sinki is Rs 50/- to Rs 60/- per kg (price index 1999-2000). The technique of gundruk and sinki production has been limited to household level and transferred from mother to daughter.

Similar products

Kimchi (Korea)
Sauerkraut (Germany)
Sunki (Japan)

Mesu

Mesu is a traditional fermented bamboo-shoot pickle with sour acidic flavour confined to limited places. The word mesu is directly derived from the Limbu dialect meaning 'me' the young bamboo-shoot, "su" soury taste. Mesu is similar to *naw-mai-dong*, fermened bamboo-shoot of Thailand.

Preparation

In the traditional method of mesu preparation, young edible shoots of bamboo are defoliated, chopped, pressed tightly into a hollow bamboo stem, covered tightly with leaves of bamboo and left to ferment at ambient termperature for 7-15 days. Fresh mesu has a shelf-life of only about a week, its pickle can be stored for a year or more (Tamang & Sarkar [59] 1996).

Socio-economy

Some rural women sell mesu at local markets. The cost of 2.50 g of fresh mesu is about Rs. 5.00 (price index 1999-2000) which is sold in the green bamboo stem.

Marchaa

Marchaa is not a food but is a mixed inocula used as a starter culture for preparation of various indigenous fermented mild alcoholic sweet-acidic food beverages such as kodo jaanr bhaate jaanr, makai jaanr, etc. Marchaa is dried round to flattended, creamy white solid ball like starter. Size of marchaa varies from 1.9 to 11.8cm in diameter with the weight ranging from 2.3 to 21.3 g (Tamang et al [61] 1996).

Synonym of marchaa

Khesung (Limbu)
Bopkha/Khabed (Rai)
Phab (Bhutia)

Buth/Thanbum (Lepcha)

Similar products

Bakhar (North-west India)
Ragi (Indonesia)
Nuruk (Korea)
Bubod (The Philippines)
Loogpang (Thailand)
Chiu-yueh (China)

Socio-economy

Price of marchaa depends on the size. Marchaa makers are confined to the Limbu caste and some Rai caste and the Lepcha of these regions. The trade is protected as a hereditary right of these castes.

Ethno-Implication

Worshipping Gods and Goddesses with alcoholic beverages such as jaanr and raksi is deeprooted ritual of the various ethnical groups of people in the Sikkim Himalaya. During marriage, traditional fermented beverage is obligatory to solemnize the ceremony for non-Brahmin Nepali, Bhutia and Lepcha, whereas fermented dairy product is essence to Brahmin Nepali, to whom drinking alcohol is a religious taboo, to solemnize marriage. Indigenous fermented foods whether alcoholic or non-alcoholic, not only provide nutrients to local diet but influence customs of the people in these regions.

NON-FERMENTED FOODS

Introduction

Non-fermented foods make the bulk of the diet of the Sikkimmese population. According to Yonjan and Tamang [64] (1998) 90.2 per cent of the diet consists of non-fermented foods, while balance 9.8 per cent consists of foods which are fermented.

Traditional Non-Fermented Foods

The traditional non-fermented foods are tabulated below (Tamang, J.P. and Tamang, N[65] 1998).

Table 5.6

Traditional Non-Fermented Foods of the Sikkim Himalaya

Product	*Ingredient*	*Nature and use*
Foods common to the Nepali		
Dheroh	Maize/millet	Cooked maize/millet; staple food
Phulaurah	Buck wheat	Deep-fried, snack
Khalo dal	Black gram	Thick dal; staple
Sidra	Dry small fish	Roasted dry fish, achar
Kwathi	Legumes	Cooked sprouted legumes; curry
Kodo ko roti	Finger millet	Baked millet flour; staple food
Phapar ko roti	Buckwheat	Baked buckwheat flour; staple food
Chambray	Rice	Soaked, fried and cooked rice; staple
Rayo sag	Brasicca spp (Leaf)	Fried vegetable; cooked leafy vegetable
Ban tarul	Dioscorea hamiltonii	Boiled/roasted
Shimrayo	Cardamine hirsuta	Boiled/roasted
Ghario Sishnu	Urtica dioica : nettle	Boiled and cooked; soup
Patle Sishnu	Laportee terminalis	Boiled and cooked; soup
Bhangrey Sishnu	Girardinia diversifolia	Boiled and cooked soup
Sauney Ningro	Diplazium polypodiodes	Fried vegetable, mix with meat/ chhurpi
Kali Ningro	Diplazium spp. Fern	Fried vegetable, mix with meat/ chhurpi
Tama	Young bamboo shoot	Boiled; curry
Kalungey Cheu	Agaricus sivaticus	Curry
Kanney Cheu	Pleurotus sajor-caju	Curry
Chhwelaa	Buff meat	Boiled, cooked meat curry
Lhakha mari	Wheat flour	Deep fried, sweet confectionery
Siltimbur	Listea cubeba	Pickle
Chinday	Pentapanax leschanaultii	Pickle
Lapsi	Choerospondix axillaries	Pickle
Dalle khorasani	Green/red chilli	Pickle
Thotne	Aconogonum molle	Pickle
Koirala	Bauhinia vereigata	Pickle
Chinping	Heracleum nepalenese	Pickle
Kukurdaine	Smilax sp	Pickle
Kabra	Ficus infecdoria	Pickle
Silam		Pickle
Vatamas ko achar	Soybean	Pickle
Philinge		Pickle
Foods common to the Bhutia		
Momo	Wheat flour/meat/ vegetable	Steamed dumpling; staple food
Thukpa/Gyathuk	Wheat noodle	Boiled, mixed with minced meat; staple
Shya phaley	Wheat flour, meat	Stuffed, fried mixed with meat
Chempo	Pork/beef	Liver cooked in sauce
Phituk	Rice/barley flour	Porridge with meat/chhurpi; staple

Phing	Vermicelli	Cooked with mushrooms, meat; soup
Khuri	Milet/buckwheat flour	Roll with meat and vegetable; snack
Phaktoo	Macroni	Macroni, cooked with meat; soup
Nya phachung	Fish	Fish baked in bamboo; side dish
Zhero	Wheat	Deep fried; Snack
Khapjay	Wheat	Deep fried; Snack
Quasi	Wheat flour	Deep-fried, cookies; snack
Tsampa	Rice/Barley	Roasted and ground
Yonten	Fried churpi	Fried in butter. Looks like staple momo.
Foods common to the Lepcha		
Kasalok	Sweet patato/yam	Cooked roots; staple
Sorongbee tulak	Rice and nettle	Cooked as porridge; staple
Cho-nyok	Rice, butter, vegetable	Cooked as porridge; staple

A CASE STUDY OF DZONGRI AND TSOKA DIETARY STRUCTURE IN HIGH ALTITUDE SETTLEMENTS

Dzongri and Tsoka and the higher reaches of Singalila Ridge consist almost entirely of subsistence agriculturists and pastoralists. Dzongri and Tsoka have long been part of the Tibetan cultural world. The eating habits of the inhabitants also are quite akin to Tibetans.

Dzongri and Tsoka food habits as listed by the villagers are as under:-

(a) **Tukpa** (*thug.pa*) (soup/stew containing water, vegetable oil, onion, salt, dried cheese, dried peas, wheat flour and optionally meat and in summer, fresh vegetables)

(b) **Tukpa namtuk** (*thug.pa.nam.thug*)(soup/stew containing water, tsampa (1.v) and salt).

(c) **Tukpa Pahtuk** (*thug.pa.phag,thug*)(soup/stew as tukpa, using butter in place of vegetable oil).

(d) **Tagyi** (*ta.qyi*) (flat unleavened bread made with wheat flour and water).

(e) **Tagyi taltak** (*ta.gyi.pral,dag*) (spherical unleavened bread, made with wheat flour and water).

(f) **Tagi skyurchuk** (*ta.gyi.sKyur.chug*)(muffin-shaped slightly leavened bread, made of wheat flour and chang (q.v.).

(g) **Paba** (*Phag.phag*)(solid doughy balls, consisting of a mixture of part-roasted wheat and barley flour, water, salt and ground dried peas).

(h) **Skiu** (*sKyu*)(small dumplings of wheat flour and water, boiled in a meat and spice broth).

(j) **Cha** (*ja*)(butter tea, made with boiling water, tea leaves, salt and butter, frequently consumed with tsampa.

(k) **Tsampa**(*tsam.pa*)(ground roasted barley grains, occasionally with ground roasted dried pea).

(l) **Kolakh** (*ko.lag*)(solid doughy balls, consisting of a mixture of tsampa(q.v.) and cha (q.v.) or tsiri(q.v.).

(m) **Chang** (*chang*)(a mild 'beer', made by fermentation of barley grains).

(n) **Tsiri** (*tsi.ri*)(watered down chang(q.v.)).

(o) **Das**(*bras*)(rice).

(p) **Dasil**(*bras,sil*)(rice, with melted butter and sugar).

a.lu(potatoes)

nyung.ma(turnip)

gobi(cabbage).

(q) **Sarktuman**(*sa.rag.tur.man*)(carrot)

(r) **Momo** (*mog.mog*)(dough ball(wheat flour), filled with a mixture of meat, onion, spices and salt, and steamed).

(s) **Oma**(*ao.ma*)(milk).

(t) **Tara**(*ta.ra*)(buttermilk)

(u) **Labo**(*la.bo*)(fresh cheese)

(v) **Churpi**(*chur.pe*)(dried cheese)

yos(roasted whole barley grains).

tsa.mig (mixture of chilli seeds, salt fresh chopped onion and radish leaves, and buttermilk; usually eaten with paba)

pha.ting (Apricots)

ka.ra (sugar)

mar (butter, from milk of drimo(female yak) or dzomo(female yak cow hybrid)

jo (yogurt)

tsong (onion)

pemar (solid doughy balls, made as kolakh(q.v.) with the addition of cheese and sugar and butter)

tso.chu (thickened broth, made by addi salt, spices, water and a little meat to small amount of paba(q.v.)).

Chub.tsos (small dumplings of wheat flour and water, boiled in a broth with spices and salt)

Conclusion

This section of the chapter on ethnobotany of the Upper Rangit

Basin inhabitants has reviewed the factors and structures of their dietary habits. The food intake of the populace also largely depends upon the economic conditions of individuals, availability of food in the environment and the relative cost, both in money and time of acquiring and then processing are different for different classes of people. The over all dietary concept however remains common.

SECTION - V
ETHNOBOTANY OF MEDICINAL PLANTS IN THE STUDY AREA

Introduction

Medicinal chemists working on plants in search of pharmacologically active principles, select for their investigations plants with known biological activity. For this information, they are indebted to nameless discoverers from long past ages. It may well be, that the knowledge was acquired empirically in the course of searching for food in the vegetable kingdom, or, as it is often supposed, that the first human beings were endowed with instincts that enabled them to recognize the curative properties of plants, and to make use of them.

This knowledge became the lore of folk medicine. In culturally highly developed countries, these medicinal plants were recorded in comprehensive herbals as in the case of Ayurveda in India. In the primitive societies including the societies of Tibet and higher Himalayas, the knowledge and use of vegetable drugs remained hidden in the hands of healers or shamans. (Albert Hoffmann[67] 1985)

Himalaya harbours about 500 species of medicinal plants and these are used for curing illness from time immemorial. Mythology reveals that the life saving drug "Sanjivani" was taken from the Himalaya by 'Hanuman' on the direction of a celestial physician 'Dhanvantari' and saved the life of Lakshman as narrated in the epic Ramayana. The ethno-cultural diversity in the Himalaya is quite pronounced, and folk medicinal practices of each group are expected to have been evolved in parallel, using the medicinal plant species. Seome of these folk medicinal practices have been institutionalised, viz, Chagpori Tibetan Medicinal practice, while a majority of them have remained unknown to the world at large. The herbal medicinal resources have been extracted for modern medicinal uses in the recent years at a large scale from all over the Himalaya. Sikkim Himalaya is very rich in the diversity of these plants (Purohit[68] 1994)

Within Sikkim and Darjeeling hills a number of institutions and individuals are engaged in research and practiced applications of the medicinal plants. These include departments of Science and Technology, Forest and Agriculture, Government of Sikkim, the Amji Clinic and STNM Hospital Gangtok. Chagpori Tibetan Medical Institute Darjeeling and Saint Alphonsus Social and Agricultural Centre. Kurseong conduct constant applied research and training in medicinal plants.

Medicinal Plant Trial by Forest Department Sikkim

The 'Minor Forest Produce' division of the Forest Department, Government of Sikkim, has been running a field trial programme over the medicinal plants and herbs of Sikkim since 1991. A total of 20 trial plots are operated in Sikkim of which the following are located in the Upper Rangit Basin in West and South districts of Sikkim (Rai and Sharma[69] 1994) :-

(a) Ralang (South District)
(b) Chingthang (West District)
(c) Yuksom Salap (West District)
(d) Yapdang Karzi (West District)
(e) Dzongri (West District)

The total trial plot area in Sikkim covers 600 hectares where different kinds of herbs have been tried. In Upper Rangit Basin the plots works commenced in years 1991 - 92 and 1992 - 93 in Dzongu, Ravangla and Yuksom.

Medicinal Plants in Dzongri, Yuksom and Bakhim

The Forest Department of Sikkim in collaboration with G.B. Pant Institute of Himalayan Environment and Development, Sikkim and The Mountain Institute, Franklin, USA undertook a research in the areas of Yuksom, Dzongri and Bakhim. Medicinal plants identified during the course of the research are listed below:-

(a) *Swertia Chirata*, Chirota (Nepali), Sunghing (Limbu), Teeka, Rungkeen (Lepcha), Kanbanwai(Gurung).

Description : Stem robust 6 to 1.5m branching tree except near the top. Leaves broadly lanceolate 10cm-x 3.8cm acute 5 nerved. Bluish green.

Habitat : Temperate Himalaya 1225m to 3000m. Mostly occurring as undergrowth or in slightly open areas. It is found in small quantity in Yuksam, on account of over exploitation in the past.

Uses : The plant is bitter, cooling, anthelmintic, antipyretic, antiperiodic, laxative, galactagogue, cures thirst, biliousness, leucoderma, inflammation, burning sensation, pain in the body, urinary discharges, ulcer, asthma, bronchitis, leucorrhoea, piles, bad taste in the mouth, good for vomiting in pregnancy, (Ayurveda).

(b) *Rubia Cordifilia,* Pandhu(Lm), Chee(Sikkimese, Bhutia), Vhyem(Lp), Rang pie (Gn).

Description : Perennial, herbaceous, climbing, roots very long, cylindrical, with a thin red bark, stem long rough, grooved, Branches scandent by means of numerous divaricate branchlets and petiole, quadrangular sometimes prickly on the angles, glabrous shining. Leaves 3.8 to 9 cm by 1.5 to 3.5 cm in whorl of 4. Ovate, acute, lower leaves larger than the upper. Scrabid 5 nerved from the base. Matured plants look yellow.

Habitat : 1000 m to 2000 m altitude. Mostly on open wasteland, in between openings of forest climbing on bushy trees, shrubs etc. Very common in Yuksom.

Uses : It produces a fast colour. In Yuksom one Gurung family has used this species to dye carpet made of lamb wool. The root is sweet, bitter, acrid, heating alexiteric, antipyretic, analgesic and anthelmintic, improves the voice and the complexion, cures diseases of uterus,, the vagina, the eye, the ear, the blood, leucoderma, drysipelas, ulcer, urinary discharges, jaundice, piles. The leaves are sweet. Increases appetite, the fruit is considered a cure to diseases of spleen.

(c) *Heraclcum nepalense*, Chumping (Np), Chumping (Sk, Bh), Sembem (Lp), Sidhumwai (Gn).

Description : 60 to 150 cm sparingly pubescent pinnate, often 7.5 to 10cm deeply lobed or pinnatifid serrate. Outer flowers of the radiant or not, Dorsal intermediate ridges hardly excurrent wing broad with a prominent marginal vitta. Flowering in rains and fruiting in early cold weather.

Habitat : Hills 1550 m, to 3600 m. Available in Yuksam, in forest as well as cultivated area.

Uses : The fruits are made into pickles, also used as an anti-typhoid treatment. Used by local people for nausea and vomiting.

(d) *Astilbe rivularis Ham*, Buro Okhati (Np), Kapoba sidda(Lm), Kharta (Sk, Bh), Tungrok supri (Lp), Bejey mwai(Gn).

Description : Stem and leaves with long fulvous baris, specially about base of the petioles. Leaflets 2.5 to 10 cm, ovate, acuminate often cordate at the base. Rachis of panicle woolly pubescent and brown villous, pedicle shoots with a small brat at the base.

Habitat : Temperate Himalaya 1200 m to 2150 m also cultivated in Yuksom by Forest Department in MFP nursery.

Uses : The rhizomes are collected and dried and preserved. Chewed as areca nut. Also boiled with milk and taken to get relief from body pain etc.

(e) *Kaempfera rotunda Linn*, Bhui Champa(Np), Samphala(Gn).

Description : Rootstock tuberous with large erect oblong, ovate-lanceolate leaves 30x7.5-10 cm usually variegated with darker and light green above and tinged purple beneath. Flower fragrant, borne 2.5 to 7.5 cm from the ground only in a crowed radical spike but only 1 to 2 opening at a time.

Uses : The tubers are used as poultice in fracture and also healing swelling. Whole plant reduced to powder and used in the form of ointment in healing fresh wounds. If taken internally it is considered to remove coagulated blood or purulent matter that may be within the body. Local people use this with mistletoe, rhizome of buro-okhati to treat fractured bones.

(f) *Saxifraga ligulata* Wall, Syn *Bergenta ligulata* bet(Np), Rimey Mwai(Gn)

Description : A perennial herb. Root stock very stout. Stem short thick fleshy, procumbent. Leaves ovate or round 5-15 cm long at the time of flowering attaining 30cm or more and turning bright red, cordate, entire fringed with short stiff hairs. Both surfaces hairy. Flowers white, pink or purple.

Habitat. Temperate Himalaya upto 3000 m. Grown in the garden and in stones walls in Yuksom.

Uses : The root is bitter and acrid, cooling laxative, analgesic, useful in tridosha, piles, tumours, strangury, vesicular, calculi, urinary discharge, heart diseases, spleen enlargement, ulcer diseases of bladder, and diseases of lungs (Ayurveda).

The root is used as a tonic in fever, diarrhoea and cough. It is also given for dysentery by the local people.

(g) *Piper longum* (Np), Kanteen(Lp), Chi Phala(Gn).

Description : Rootstock erect, thick, jointed branched, stem numerous, Leaves numerous 6-9cm, lower ones broadly ovate, upper ones oblong oval cordate at base. Dark green and shining above, pale and dull beneath.

Habitat : Grows upto 600 m in the foothills. Found on rock and stones in Yuksom.

Uses : The root is pungent, healing, stomach, laxative, anthelmintic, carminative, improves appetite, useful in bronchitis, abdominal pains, diseases of spleen, tumours etc. The ripe fruit is sweetish, pungent, healing, stomachic, aphrodisiac, alternative, laxative, anti-diarrhoeatic, anti-dysenteric, useful in bronchitis, abdominal complaints, fever, leucoderma, urinary discharges, tumours, piles, diseases of spleen inflammation, leprosy, insomnia, jaundice, hiccough, tuberculosis glands etc.

(h) *Rhodoendron arbocum*, Gurans (Np), Thok Pheklapung (Lm) Ado Mitho (Sk, Bh), Olgya Taha (Gn).

Description : A small evergreen tree which attains girth upto 1.5 m and height of 8m. Bark pinkish brown, rough exfoliating in thin flakes. Blaze white or pinkish. Young shoot clothed with while scale. Leaves 7.5 to 15 cm by 3.5 to 5.0 cm crowded towards the end elliptic oblong acute at both glabrous above, rusty tomentose or with a dense silvery film of minute hairs. Flowers in February in warmer place and March-April in higher altitude. Colour crimson or pink.

Habitat : Outer Himalayan 1500 m to 3300 m. It is also cultivated in Yuksom.

Uses : Wood is made into Khukuri handle and pack saddles. Also good fuel. Flower is sour collected and dried for using as a cure for dysentry. Many preserve the flower petals as deobstruent of fish spine. The flower petals are also brewed into a local wine for using in stomach trouble.

(j) *Oraxylum Indicum*,Totala(Np), Paksam(Sk, Bh), Poksum(Lp), Koko Mhendo (Tm)

Description : A small soft wooded deciduous tree bark silvery grey or brown when matured, corky. On stem and branches there are distinct heart shaped scar of fallen leaf pinnate. Leaflets ovate, acuminate blade 10-12 cm x 7-10cm. Petiole short Flowers in May-June. Fruits in winter (December-January) are prominent when the tree is leaflets. Long broad sword shaped with many light wringed seeds inside in 2 septa.

Habitat : Throughout India in sub-Himalaya it grows upto 1000 m. This is not seen in Yuksom but can be grown.

Uses : The flowers are edible as vegetable. Root bark is acrid, cooking, aphrodisiac tonic, increases appetite useful in biliousness, fever, bronchitis, intestinal worms, vomiting dysentery, asthma, inflammation, anal troubles. The stem bark is also extensively collected to be used as lever tonic, endangering the existence of this species. The fruit is an expectorant and improves appetite, also useful in leucoderma under Unani Medicine. The seed is used extensively by Buddhists as flowers. The local people used the seed for curing pneumonia. The dry food is also in great demand as decorative.

(k) *Plantago major*, Nasey Jhar (Np).

Description : A perennial herb with an erect stout root stock. Leaves alternate, radical 2.5 to 12.5 cm long of variable width, ovate or oblong, obtuse or sub acute entire or toothed nearly glabrous tapering and decurrent into the petiole. 3 to 7 nerved petiole upto 10 cm sheathing at the base. Flowers scattered or crowded in long slender rather lax spikes 5-15 cm long.

Habitat: This was found at 1750m height in the MFP Nursery of Forest Department in Yuksom.

Uses: The plant is useful in rheumatism and gripping pain of bowels. The leaves and root are astringent and used in fever. Seeds useful in dysentery.

(l) *Polygala arillata*, Marcha (Np).

Description : An erect pubescent or nearly glabrous shrubs 1 to 2.3 m high. Leaves 1015 cm x 2 to 6 cm lanceolate or elliptic oblong acuminate petiole 10 to 12 mm. Flower usually in panicled raceme 2 lower sepais large like wings, ovate and red purple, corolla yellow, capsule 12 to 20 mm broad-reniform, rather fleshy.

Habitat: Yuksom 600m to 1800m.

Uses: The root is used to make yeast.

(m) *Solanum xanthocarpum* Jungli bihi (Np).

Description: A very prickly diffuse bright green perennial herb, stem somewhat zigzag. Younger branches clothed with dese stellate tomentum. Prickles straight yellow. Leaves 5-10 by 2.5 to 6 cm ovate or elliptic. Stellately hairy on both sides. Glabrous when old. Base usually rounded and unequal sided. Petiole 1.3 to 2.5 cm long stellately hairy and prickly.

Habitat : This was found on the road to Dubdi Monastery.

Uses : Root is useful in bronchitis, asthma, fever, strangury, lumbago, pains, piles, thirst, urinary concretions and heart diseases. The fruit improves appetite, good in heart diseases, asthma, fever and anthelmintic. Locally the seeds are burnt and its smoke applied for toothache. Believed to remove the germs and thereby relief from the toothache.

(n) *Aremisia vulgaris*, Titepate (Np), Taknyel(Lp), Pati Kamba (Gn).

Description: Perennial shrub, aromatic 0.6 to 2.4 m high, paniculately branched. Lower leave 5.10 by 2.5 to 5 cm petiole. Ovate in outline. Upper leaves smaller 3 feet or entire lanceolate.

Habitat: In open areas of forest and waste land in Yuksom. Also along road side.

Uses: It is valuable treatment for stomach complaints and used deobstruent and antispasmodic. It is prescribed in infusion and electuary in cases of obstructed menses and hysteria. Its fomentation is applied on skin diseases and foul uncer. The juice is applied on the head of children to prevent convulsion. The leaves and tops are given in nervous and spasmodic affections connected with debility in asthma and diseases of brain. The local practitioner of herbal mediciner of Dosthang village applies hot fomentation of black stemmed. Titepate called kala Titepate for treating cataract on eyes.

Local Ethnic Practices

Though different ethnic groups inhabit the Upper Rangit Basin, the different systems of herbal package found here are broadly of two types - The Nepalese Systems and the Tibetan System. The Lepchas, though found to be using a number of herbs, are more inclined towards the use of animals as medicinal resources (Rai and Sharma[70] 1994).

Nepali System of Herbal

Deeply seated in its antiquity but very thinly documented the Nepali system of herbal survives today as Jaributi or just simply as *Pahaday dabai*. The practice of herbal medicine has in its true nature, not yet attained a system status in itself as in the case with Tibetan Medicine. Collection of the jaributi herbs, its use and affecting a sale in the market can hardly be acclaimed as a system but nevertheless a small-scale system definitely is found operating. It is certain that the advent of allopathic 'pill culture' have rested the glory and to a greater extent the usage of jaributi in the hills, however, romance and attraction still lingers and at every haat bazar a few jaributi peddlers still manage a brisk business.

Most of the herbs used in the system are similar to the Tibetan system. As because it is not an organised system the jaributi system has to bring up the several disjointed threads together to get a proper status. At present the system survives at the peddler level and at some villages where each roof keeps a few jaributi at its backyard for future use. Some of the major plants used in the jaributi, and 'recipes' are given below :

Name	*Usage*
Harchur	Bone fracture
Harra	Cough
Barra	Cough
Jethi madhu	Pneumonia, cough
Kuthki	Shivering fever
Majhiphal	Tonsillitis (with warm water)
Chhatiwan (bark)	Diabetes
Phachyeng	Jaundice (with brandy)
Haledo	(with Phachyeng)Burns
Pangra (kernel)	Mumps
Blimbira (root)	Worms in children
Jatamashi	Cuts
Panchaunlay	Bodyache, heart conditions
Padamchal	Blood clot/swellings
Pakhanbhed	Oral inflammation/infection
Nagesori	inflammation & septic conditions
Kali ningro (root)	Dysentery
Rach bita	Colic conditions
Indrajau	Recurring fever & vomiting

Lepcha System

Interestingly enough it seems that no Lepcha herbal system exists at the present. This is in outright contrast to the fact that almost all the plants of the region are identified, named and thus known to Lepchas from a vast period in history. The use of animals as form of 'zootherapy' is quite popular as well as pronounced with the Lepchas in which all types, from bear to fox are used. (Rai and Sharma[71] 1994).

Tibetan herbal system

Tibetan medicines hardly ever use a single plant as a prescription. Normally a composition is made out from different medicinal herbs to treat a certain disease. It is out of the singular belief that the process would derive the optimal medicinal advantage and/or to identify any toxins therein. The Tibetan medical practitioner in Ravangla has a brisk business in the Tibetan Refugee Settlement and Ravangla market.

Tibetan medical prescription usually comes in the form of powder and different form of extracts, much in contrast to medicine of Nepalese where the prescription normally is unprocessed plant product and comes in all of its natural shape, size, colour and form (such as bark, twigs, fruits, leaves, seeds, etc).

Spirituality is undoubtedly the mainstay of Tibetan Medicine of which Chogpori Tibetan Medical Institute is a perfect example. The system with its deep roots in ancient Tibet where spirituality dominated the scene for long and socio-economic ramification was attached with the Dharma, the medicine system was not an exception, it being originating from the pivotal point namely the sacred Potala and various monasteries. But the 'general practitioner' of Tibetan Medicine at present relies more on both, plants and spirituality.

It is known that usually in the Tibetan system plants are mixed together to make a prescription compound and so act in a synergistic way, but information on this aspect so far is eluding. Also, the system uses, apart from herbs, different physical additives (gold pearls, minerals, etc) in various prescriptions. (Rai and Sharma[72] 1994).

Recommendations/Suggestions

Expertise for Conservation

No single sector, private or public can undertake the conservation of medical plants alone. The job requires a team effort involving a wide range of disciplines at institutions (IUCN, WHO and WWF[73] 1993).

According to IUCN, the experts most needed for a programme of conservation and sustainable utilization of medicinal plants, would include the following :-

(a) *Agronomists*	: To improve techniques for cultivating medicinal plants.
(b) *Conservation Campaigners*	: To persuade the public of the need to conserve medicinal plants.
(c) *Ecologists*	: To understand the ecosystems in which medicinal plants grow.
(d) *Ethnobotanists*	: To identity the use of plants as medicines in traditional societies.
(e) *Health Policy - makers*	: To include conservation and utilization of medicinal plants in their policy and planning.
(f) *Horticulturists*	: To cultivate the medicinal plants.
(g) *Legal Experts*	: To develop effective legal mechanisms that ensure that collection of medicinal plants is at levels that are sustainable.
(h) *Park Managers*	: To conserve medicinal plants within their and reserves.

(j) *Park Planners*	: To ensure the park and reserve system contains he maximum diversity of medicinal plants.
(k) *Pharmacognosists*	: To study the application of medicinal plants.
(l) *Plant Breeders*	: To breed improved strains of medicinal plants for cultivation.
(m) *Plant Genetic Resource*	: To assess and map the genetic variation specially of medicinal plants.
(n) *Plant Pathologists*	: To protect the cultivated medicinal plants from pests and diseases without using dangerous chemicals.
(o) *Religious Leaders*	: To promote a respect for nature.
(p) *Resource economists*	: To evaluate the patterns of uses and the economic value of medicinal plants.
(q) *Seed Biologists*	: To understand the germination and storage requirement of the seeds of different medicinal plants.
(r) *Taxonomists*	: To identify the medicinal plants accurately.
(s) *Traditional Health*	: To provide information on the uses and practitioners availability of medicinal plants.

Policy Planning. (IUCN, WHO and WWF[74] 1993) suggested following measures for policy planning :-

(a) The state should identify and support one or more institutions to plan, coordinate and implement ethnobotanical surveys.

(b) The selected institutions should implement a state programme of surveys on the use of plants for medicinal purposes in traditional societies.

(c) The data on ethnobotany should be catalogued and analysed but only disseminated in such a way that the communities providing the data would receive benefits from any commercial use of the information.

(d) The ministry of Health should incorporate proven traditional remedies into state programmes of Primary Health Care.

(e) Traditional Health Practitioners should constitute themselves into state and national bodies.

(f) The State should have an adequate herbarium with a botanical library, so that plant material can be adequately identified and stored within the state and a cadre of well trained botanists able to identify plants, who can staff the herbarium and other botanical institutes and departments.

(g) Botanical institutions should make a catalogue of all the plant species used for medicinal purposes in the state:

(i) Its scientific (Latin) name, and its vernacular name.

(ii) Its geographical distribution.

(iii) Its scarcity or abundance, population size and conservation status

(iv) Its habitat.

(v) Description of the part of the plant used (e. g. stem or leaf) and how collected.

(vi) Its use in medicine, traditional or otherwise.

(vii) Its occurrence in protected areas.

(viii) Whether cultivated or not, and the availability and location of ex situ germplasm.

Conclusion

Good health and conservation of diversity of the medicinal plants would help the traditional ethnic folks retain their generations old traditional ecological knowledge. It is the prime duty of the state government and the educated intellectuals to help retain this important aspect.

SECTION - VI
NON - TIMBER FOREST PRODUCTS IN YUKSOM : A SURVEY

General Description of Yuksom

Location and Terrain. Yuksom is bound on its NW and SW by Rathang Chu.

Its lowest altitude at the Rathang Chụ is about 1000 m which goes upto 3000m in the forest situated above Yuksom. The habitation is mostly on the spurs and gentle slopes. There is one village called Ramjithang which in Lepcha means the deposit of landslide. The area is steep to precipitous except the spurs and the main Yuksom, Khyontey and Norbugang. Most of the natural forests up to 1800m are converted into settlements. Easier slopes wherever available are under cultivation, and steeper, rocky areas covered with either secondary growth or bamboos. Areas where paddy used to be grown once are replaced with cardamom under *Alnus* (alder) trees of about 25 years in age conversion. This was done on account of poor paddy production, On the South west side above the road to Pelling from Yuksom, there has been large scale landslide caused by construction of approach road to the Rathong Dam site. The forest around Dubdi Monastery and above 1800m constitute *Castanopsis*, Oaks, Laurel, Maple, Rhododendron with thick under growth of Maling bamboo, *Rebus, Daphne, Symplocos, Eurya, Viburnum* ferns, etc.

Historical Importance. Yuksom is regarded as the first capital of Sikkim. From Yuksom if one walks for 15 minutes he will be in Norbugang where one can see a Royal sear in stone *masonry* where the crowning ceremony of the first Chogyal, *Dharma Raja* of Sikkim took place under guidance of Lhotsum Chhenbo a Tibetan priest from Tibet. It has also the distinction of having the oldest monastery, Dubdi.

Folk Belief. Yuksom is endowed with rich natural beauties. The Famrang Fall on the entry from Tashiding and view of Kanchendzonga while entering from Pelling side are attractive sites for outsiders as well as the locals. The snowclad Pandim mountains on its North looks like the shining crown on its head. There are two holy lakes, Kathok within Yuksom and another Khecheopalri some distance away from Yuksom which sanctifies the whole region. There is belief of the people that the Khechopalri was once in Yuksom itself as the female consort of Kathok but on account of neglect and pollution by the people it shifted to its present position. The depression with opening is still there on South West of Yuksom Bazaar indicating the old site of its existence. How far this belief if true is not easy to ascertain. It remains a folklore.

Besides these, rich forest surrounds the upper boundary which goes right up to alpine region. These have been the source of clean water in the various rivers and streams that flow through Yuksom. Under the cool shades of these forests and in the folds of these Himalayan mountains rich treasure of forest products other than timber are available that are so important to sustain the life system. They consist of bamboos, broomstick grass, fruit trees, buds, leaves, flowers, valuable medicinal plants, mushrooms, ferns, natural decoratives, etc. There are a large number of plants in Yuksom that are important for medicines which are used by local people even to cure serious diseases, sprains, cuts and fractures etc.

The language specialists deserve special thanks by giving the names of different species in their own languages. The names of each language specialist is given against the following languages :- Limbu (Lm) Sri Man Bahadur Subba of Topsing, Sri H.B. Khamdak, Limbu teacher of Yuksom High School, Sri Sanchabir Subba of Khyongtey village. Bhutia (SkBh) :- Sri Sonam Wangdi of Yuksom. Lepcha (Lp) :- Sri Sonam Tshering Lepcha of Yuksom and Sri Motey Lepcha of Khyongtey village. Gurung (Gn) :- Megraj Gurung, President Simmik Gurung (Tamu) Youth Association, . Sherpa (Sh) Phurbu Sehrpa. Shri SD Lama (IPS) provided most researched data as presented in this section of the chapter.

Sikkim Biodiversity and Ecotourism receives support from the collaborating institutions and communities, donations and grants obtained from the United States Agency for International Development including support from the (USAID) Biodiversity Conservation Network (a programme of the Biodiversity Support Program, a consortium of World Wildlife Fund, the nature conservancy and the World Resources Institute), and the Bureau of Humanitarian Response. This project is a collaboration of the Mountain Institute (USA), G.B. Pant Institute of Himalayan Environment and Development, the Travel Agents Association of Sikkim and the Green Circle.

Non-Timber Forest Products Uses

In the past when communication was difficult, people depended for their requirement of food, fuel, fibre, fruits and medicines on these products called Non Timber Forest Products (NTFP). Bamboos and canes were used commonly by local people for construction of bridges across various hill streams as the only means of communication. The section aims in bringing out the various non timber forest products, identify them with the help of local people in their own languages and its underlying importance as essential commodities of our life system, so that awareness is created for their propagation and preservation. A total of 72 NTFPs of Yuksom are described under the following headings:

1. Bamboos

Though there are more than 10 varieties of bamboo in Yuksom, of which descriptions of only 8 varieties of economic importance are given here. Most of the bamboos are cultivated except Niglo, Gopey and Maling. The Pareng bamboo though cultivated, is also found in village forest and government forest. Some carrier items made of out of bamboo stripes are shown at Figure 5.4.

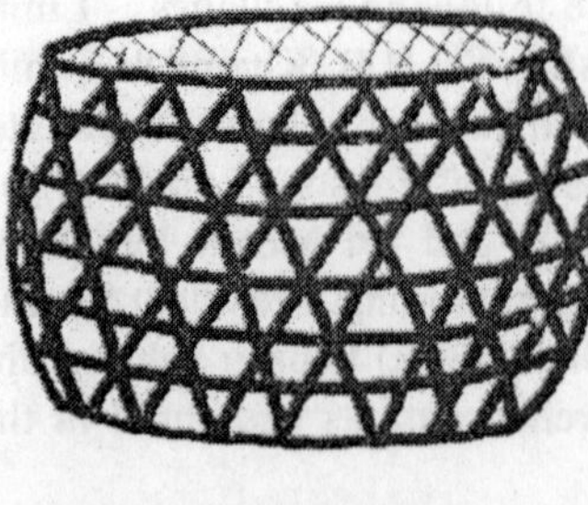

Bamboo-made *Septu*

Wood-made *Septu*

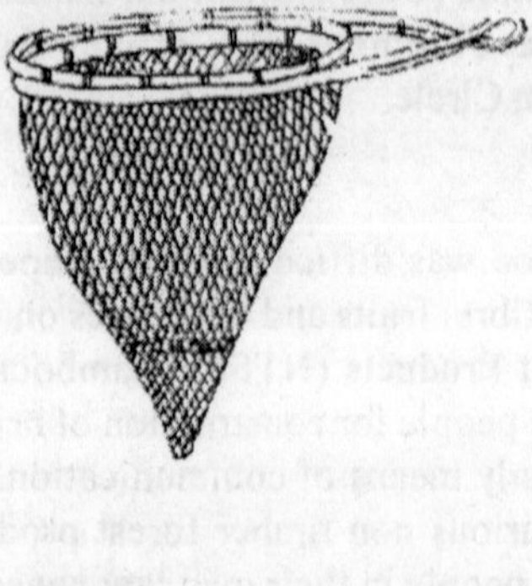

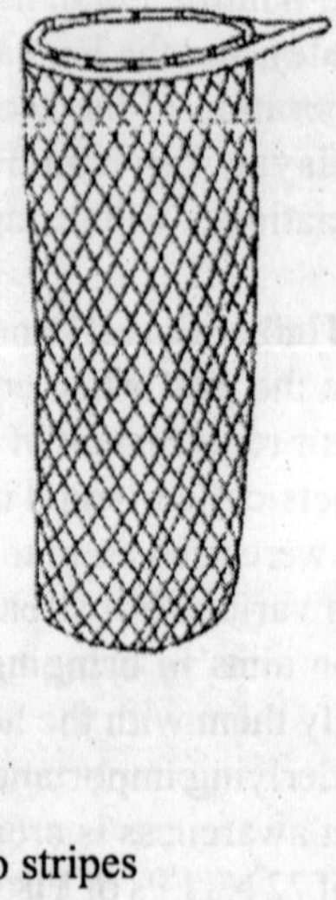

Chappani made-up of bamboo stripes

Fig. 5.4

Non-Timber Forest Produce, Sikkim

2. Amliso (Broom Grass)

Though Government had also supported its extension through social forestry scheme and people also had necessary will to plant this grass as cash crop for its broomstick, but its slide in price from 1997's harvest of January-February has seriously affected the interest of the growers. However, the value of this grass as cattle fodder and soil stabiliser should not be lost sight of

3. Fruits, Berries etc.

Trees in forest as well as on own land offer a variety of fruits not only to human beings as nutritional diet but they are good source of food for wild life as well as birds. Some of them happen to be good medicines also. Some 23 fruits bearing trees are described in this report.

4. Buds-Young Shoots and Leaves

There are a number of trees in forest whose young shoots, leaves and flowers are eaten as vegetable or made into pickle. Some of them also are sources of medicine. Six plants are described under this section.

5. Medicinal Plants

Although most of plants available in Yuksom and around are used by local people practitioner (Bijuwa or Baidya) of herbal medicines, only 13 important plants of Yuksom which either are used in local herbal medicine or some of their efficacious uses are not known. However, the idea of giving detailed use of these plants under Ayurved or Unani is to create awareness about these valuable species about which many are unaware. The details of medicinal plants are enumerated in previous section of this chapter.

6. Decoratives

Natural forests are source of varieties of natural attractive materials in the form of decoratives. Roots of plants, dry flower, capsules, dry, mushrooms cones of conifers, leaves of fern, fern shoots, seeds of 'totala' all go to form decoratives of all designs and imaginations. In Yuksom nothing is done in this field.

7. Mushrooms

The local people collect and eat about 11 varieties of mushrooms in Yuksom. Except one, all are found on natural dead wood during rains. Only eight varieties are described under this chapter.

8. Diplazium (Ningro)

Good quantity of Diplazium (Ningo) comes to market during rains. Many directly collect and take the young shoots of fern as good vegetable. Three varieties of Ningros are described here. They form important vegetables for the local people.

9. Yams

Forest have a number of dioscoreas which provide food to people through their yams. Only one variety i.e. Dioscorea (Ban Tarul) is collected from forest. The other dioscorea that is D. bulbifera (Ghar Tarul) is cultivated in lower zones upto 1000m. The first one only is described.

10. Fibres

Argeli (Edgeworthia) and Lokhtey (Daphne) are two plants widely used by locals for making fibre materials. These plants especially Argeli grows abundantly in and around Yuksom.

DESCRIPTION. HABITAT AND NTFP USES

Bamboos

(a) *Dendrocalamus sikkimensis* - Bhalu Bans, Dhiri Bans, Dhungra Bans (Np), Patek pu (Lp), Hang bha (Lm), dhungsing (Bh), Patep (Sp), Ngugma (Tb), Rein Rama (Gu).

Description : Very large and handsome bamboo growing in the lower hill valley upto 1800 m. Density tufted, evergreen culns erect 25-30 m high. Internodes 35-50 cm, diameter 15-20 cm.

Habitat : It is available in Yuksom upto 1800m.

Uses : The local people use it for carrying water, churning milk to make butter milching bucket etc. The Sherpas call it Tolang. It is also split and used for making walls, partition of houses and shades, floors and fencing of garden. Pictures drawn on small pieces can form good decorative pieces. Local artisans could be trained on this art which may become good source of income through sale to tourists.

(b) *Dendrocalamus hamiltonii.* Tama or Choya bans (Np), Yekyot-pao (Lp), Dheybha (Lm), Chhasing (Bh) Rie(Gu).

Description : It is the most common tufted forest bamboo. The culms grow upto 24m high with 10-18cm diameter. Internodes are 30-50cm. The culms have thick walls when old. The lower parts are glaubose.

Habitat : The bamboo is found naturally occurring in village forest and also cultivated in the village upto 1730m in Khoyngtey above Yuksom bazaar.

Uses : Its bigger culms are used as water-conduit, water carrying vessels called Dhiri. Matured culms are used for construction of houses for structural purpose and as fencing material. Its bigger internodes are made into beer jugs called Tongba. Young shoots are extensively collected in July-August and eaten as vegetable or made into pickles.

Semi -fermented *tama* is sold in the market as "meso". It can be preserved and used in off season to add flavour and taste to curry. Very good bending cane is extracted from outer layer of this bamboo which is locally called *choya*, hence, the name *choya bans*. This also forms good raw material for paper pulp. Also these bamboos can be made into decorative pieces. The leaves form good fodder for cattle.

(c) *Dendrocalamus hookeri*. Chilley Bans (Np), Pata (Lp), Rein Pochoarey (Gn).

Description : Tufted culms 18 to 20m, internodes 45 to 50 cm, diameter 7.5-12cm. Stem bright green smooth, young ones covered with brown hairy sheaths gradually narrowed from a broad based to a truncate at top.

Habitat : This bamboo is cultivated quite extensively in Yuksom for use in various constructional purposes upto 1750m.

Uses : Used in wall and roof structures, fencing support of squash etc. The shoot is bitter hence not eaten. However, some villagers boil and throw the water and eat the shoot as vegetable, like that of D.hamiltonii.

(d) *Arundinaria hookeriana*. Pareng (Np), Pagjiok po (Lp), Parangma (Lm), Paran (SK, Bh) Chyoba rien (Gn).

Description : Medium sized tufted bamboos with glaucous stem, Culms 4 to 10m, internodes 20-30cm, diameter 20 -40mm.

Habitat : Between 1200m to 2100m in village forest as well as reserved forest in Yuksom.

Uses: It is used for making mats, roofing materials for houses shades, baskets of various sizes winnowing tray etc. Roofs are also built with short splits of this bamboo. The shoots from the villagers own field are taken as vegetable.

(e) *Arundinaria intermedia*. Niglo (Np), Parmiok (Lp), Thing mey (Lm), Barming (Sk, Bh), Thing ningma (Sh), Reen chetowa (Gu).

Description : Small thin bamboo. Height 2.5-3.5m internode 12-25 mm.

Habitat : They are found on dry slopes of village forest between 1200 m to 2100 m.

Uses : It is used for making mats, baskets, house tieing by its cane etc.

(f) *Arundinaria racemosa*, Malingo (Np), Mikla (Lm), Phiung (Sk, Bh), Ningma (Sh), Phylum, Pervok (Lp), Rien chayba (Gn).

Description : It is the most common bamboo of the forest of Yuksom, grows gregariously, single stemmed, 3-l0 m high with internodes 30-35 cm. Its diameter is 15-40 mm.

Habitat : 1850 m upward upto 2750 m in forest of Yuksom.

Uses : It is extensively used for making roofing mats, basket fencing, walking stick and flute. The splitted pieces are used in reinforcement of wall. Villagers are seen collecting this bamboo in two bundles of 30 each with 3.0 m length for above purpose from forest.

(g) *Bamboosa nutans*. Mala bans, Mahlu (Np), Multapha (Lm), Thuysing (Sk,Bh) Lee malu po (Lp), Malarien (Gn).

Description : Large, graceful bamboo, stem solitary grows upto a height of 6-20 m. internode 30-60 cm, 5-12 cm diameter, wall thick, 300-1550 m.

Habitat : It is cultivated extensively below the road to Pelling from Yuksom, at an altitude of 1300 mm.

Uses : Since this bamboo has hardwood, it is used for structural purposes such as poles, support for shuttering in building construction. Many split this bamboo and use it as reinforcement in cement or earth plastered houses in the villages. Also used for making basket. Buddhists use this bamboo as prayer flagstaffs.

(h) *Cephalostachium capitatum*. Gopey bans (Np), Yuem (Lm), Pangrim (Sk, Bh), Peyong (Lp), Rein Kayey (Gn).

Description : It is gregarious, densely branching, more or less scandent, culms 3, 5-10 m, yellow internodes 60-90 long, 25-40 mm diameter. Leaves pale green ovate-lanceolate blade 15-20 cm.

Habitat : It has the highest range of occurrence starting at 600 to 2400 m in Sikkim. Plants are found scattered in steep slopes in village forest in lower zone and in Reserved Forest at higher reaches.

Uses: The leaves are excellent fodder. Culms are used by Lepchas for making bows and arrows. The internode also makes good flutes. The locals use the bamboo for making straw for drinking local beer (Chhang).

Bamboo Flowering: almost all the species of bamboo flower gregariously followed by its death. The profuse seeds that fall after flowering regenerates in the entire area provided it is not damaged by fire. Different bamboos flower at various intervals varying between 12 years to 25 years.

Broom Grass

(a) *Thvsanolacna maxima* - Amliso (Np), Setlu (Lm), Passeam (Sk,Bh)Passeor (Lp), Chakusey (Sh), Burdhun (Gn).

Description : A very large glabrous grass growing upto 1-4 m high, diameter of the stem upto 12 mm. Leaves broad flat 60-40 cm by 5-10 cm coriaceous, tapering to a fine point. It has typical dented mark across the leaf blade at 2/3rd its length. The panicles which are 0.4 - 1.2 m long from the broomstick.

Habitat : Grows in the sub-tropical Himalayas from plains to 2000 m altitude. It is extensively planted in the hills in waste land and as inter-crop in agroforestry system and also on edges of terraces. In lower villages the amliso is cultivated for the last 4-5 years in small areas and on edges of terraces with government incentives.

Uses : The inflorescence produces the soft broom for cleaning the floors. The leaves serve as very good fodder for the cattle especially during the lean period. The stick is used as fuel wood after drying.

The following table gives the per hectare production of broom stick, green fodder and drystick as per study conducted in Karmat village at 600 m altitude in Darjeeling District Lama[75] (1998) :

Age (year)	*Dry Broom (kg)*	*App Value (Rs)*	*Green fodder (kgs)*	*App value (Rs)*	*Dry stick (kg)*	*App value (Rs)*	*Per value (Rs)*	*Fodder for cattle*
1.5	676	6780	7040	3520	313	313	10593	141
2.5	1099	10990	12400	6200	708	708	17898	248
3.5	1553	15530	11800	5900	605	605	22035	236
4.5	882	8820	9640	4820	388	388	14028	193
5.5	595	5950	7240	3620	215	215	9785	145

From the above table it transpires that with every year the crop production starts falling from 3.5 years onwards. This calls for application of fertilizer and replacement of old clumps by new vigorous seedlings.

Fruit Bearing Trees

(a) *Castanoosis hvstrix*. Patte Katus (Np), Chestnut (Eng), Seybheuengma (Lm), Sumsi (Sk, Bh), Kushukung (Lp), Kasing (Gn) Kasyopot (Sh).

Description: Generally medium sized tree rarely large size, young shoots pubescent, leaves thinly coriaceous entire sometime serrate near the apex long acuminate. Blade 10-20 cm. Flower March-April, Fruits October-December.

Habitat : Occurs quite gregariously in upper hill forest of Yuksom between 1800 m to 2450 m altitude.

Uses: The seeds are available at an interval of 3-4 years and are collected by villagers either for sale or for consumption. The dry leaves are collected for making bedding of cattle and form very good compost.

(b) *Castanoosis tribuloids*. Musre Katus (Np), Khechingsey (Lm), Sokey kap (Sk,Bh), Kushu nok (Lp), Seshy (Sh).

Description : This is a middle sized rarely large tree. Stem rough, irregular, bark grayish. Leaves thinly coriaceaous, entire sometime serrate near the apex, lanceolate, long acuminate. Blade 10-20 cm upper side glabrous, underside pale or reddish sometimes pubescent. Spikes are upto 25 cm long spikes upto 10mm long, usually on short ridges, often stellate on branches. Nuts 1-3 glabrous when mature, cotyledons ruminated flowers in March-April Fruits in October-December.

Habitat : It is found gregariously growing near the Forest Rest House in Yuksom and around and below Dubdi Monastery.

Uses: Same as C. hystrix

(c) *Quercus pachvohvlla*, Sungurey Katus (Np), Mikthumba (Lm), Den (Sk, Bh), Songpang-kung, Sarob-kung, Sri-Kung (Lp) Oak (Eng).

Description : Large sized trees with spreading crown. Leaves entire, acuminate, glabrous above with minute stellate hairs. Leaves 12 to 20 cm long. Flower in June-July. Fruits in November in clusters with thick cover, covering nuts more than half it portion.

Habitat: Grows between 1830 to 3000 m along with patley Katus, Buk, Maple and laurel and in upper zone with rhododendron.

Uses: The nuts are found and dried and made into flour out of which bread used to be made in the past as per version of old people. Bears enjoy the acorns for which they break branches and make temporary abode at the top. The bark and acorns are used as astringents in Sikkim. Similar use is made from the bark and acorns of Quercus semicarpifolia another oak that occurs in the hills between 1830 to 2450 m. Its leaves are broader and silvery underneath with prominent lateral nerves.

(d) *Junglans regia*, Okhar (Np), Khese (Lm), Taga (Sk,Bh), Kalkung, Kowal (Lp), Kotachi, Kolpot (Sh), Kado (Tam).

Description: Large deciduous aromatic tree, young shoots tomentose. Bark grey deeply fissured vertically. Blaze red brown. Leaves imparipinnate, leaflets 5-13 sub-sessile, opposite or nearly so, usually entire, serrate in seeding stage. Flower March-April, fruits September-November with hard shells.

Habitat: Middle hill forest between 1000m to 2000m mostly in depression also cultivated in private land and cardamom fields in Yoksom.

Uses: (i) Bark - used as an anthelminitic and detergent.

(ii) Leaves - Astringent and tonic. Its decoction is specific in strumous sores. Also anthelmintic bark and leaves are used in herpes, eczema, scrofula and syphilis. Externally the infusion is used as an application for skin eruptions, ulcers, etc.

(iii) Vinegar of pickled young fruits make effective gargle for sore throats. The oil from kernel is applied externally to cure troublesome skin diseases of leprous type. Kernels are also used by some to cure heart bum, colic and dysentery. In Yuksom fruits are sold in the market.

(e) *Machilus odoratissima*. Lalikaulo (Np) Sapsing (Lm), Phamlex (Sk, Bh), Rohunkung (Lp), Olglya, Kansindu (On).

Description : Medium sized tree. Bark rough brown, thin blaze reddish while inside, Leaves glabrous, dark green shining above, pale beneath, lanceolate-ovate or slightly ovate. Midrib very prominent on the under surface, aromatic scented. Flower March-May, sepals glabrous, fruit October-December.

Habitat: Hill forests of Sikkim between 1500 m to 2150 m altitude in association with oak, maple, magnolia, etc.

Uses : Mature fruits are collected from ground and the outer pulp eaten like that of Machilus edulis. Leaves are good fodder for cattle, but it should not be allowed to be lopped as this will hamper the growth of the tree.

(f) *Machilus edulis*. Lapche Kawla (Np), Phumsey (Lm), Pamshi (Sk, Bh), Phumkung (Lp).

Description: Large sized tree. Bark brown with prominent lenticels, bole cylindrical. Leaves glabrous above tomentose beneath Blade 12-30 cm. Young leaves densely pubescent. Main veins prominent beneath. Flowers May, fruit January -March, size upto 3.5 cm diameter. Fruiting at 3-4 years interval.

Habitat : Middle to upper hills 1220 m to 2400 m. Also cultivated in Yuksom.

Uses: Fruits are cured for some days in earthen pits and the pulp eaten after kneading. They are also sold in the market. Leaves are good fodder for cattle.

(g) *Quercus lineata*. Phalant (Np), Srising (Sk,Bh), Sri Kung (Lp).

Description: Large sized trees. Bark grey brown with rough spots. Leaves ovate lanceolate serrulate on the upper half glaucous beneath. Capsule shallow 25 mm diameter surrounded by thin lamellae. Flower April-May, fruits September-October.

Habitat: Upper hill forest of Sikkim 1850-2700 mm. Also in forest of Yuksom.

Uses: Acorns good food for bears.

(h) *Prunus nepalensis*. Arupate (Np), Khamkin (Sk, Bh), Tukporuynok (Lp).

Description: Bark dark brown with horizontal lenticels. Leaves with acute or rounded base. Serrate margin, acuminate young leaves pinkish to light green appearing in February-March, fruit May-June.

Habitat: Yuksom forest above 1800 m.

Uses: The outer thin pulp turns black when ripe and are edible. Birds also devour the fruits

(j) Cinnamomum impressinervium, Sis (Np), Slothkung (Lp).

Description: Medium sized tree. Bark brown, blaze scented. Leaves opposite, triplinerved, shining above pale beneath small, veins deeply impressed on the upper side. Aromatic small when crushed in July, fruits in December.

Habitat: From middle to upper hill growing in second storey. Occurring between 1220 m-1830 m in Yuksom forest.

Uses: Fruits mature in December. They are collected from ground and the outer bark rubbed and washed. The seeds are fried and eaten. The ripe fruits attracts birds like maina, parakeet, grey pigeon, green pigeon etc. to the hills during winter.

(k) *Spondias axillaris*, Lapsi (Np), Silotkung (Lp), Phindruksa (Lm), Lingsu (Sk, Bh), Phalaan (Gn).

Description: A tall evergreen tree. Bark brown or reddish peeling off in long flakes. Leaflets 6-8 pairs, opposite, or nearly so, ovate lanceolate, acuminate, in old trees entire in young trees, serrate 7-12 cm long, petiole 6 mm long. Flowers in April, fruit ripens in December. Drupes around 25-30 mm long yellow when ripe.

Habitat: In Yuksom it grows between 800-1400 m.

Uses: The ripe fruits are eaten as such or made into pickle. Out of pulp 'Titora' a dry pickle is also made. The fruits are sold in the market in Yuksom.

(l) *Elaeocarpus lanceaefaius*, Bhadrase(Np), Pego lechey(Lm), Khasyamkeon (Sk, Bh), Zeerkung, Sakyu kung (Lp).

Description: Large sized tree with dark brown bark. Leaves lanceolate, thinly coriaceous narrowed into a short petiole. Secondary nerves 6-12 pairs joined by intramarginal veins. Old leaves red. Flower spikes pendent. Flower April-May, fruit September, ellipsoid 30mm long.

Habitat: Yuksom forest above 1830m to 2450 m

Uses: The fruits are edible. The stones are often collected from the ground left by deer after regurgitation or after the outer mesocarp is rotten. The kernels are extracted by breaking these stones and eaten.

(m) *Bassi butvracea.* Chewri(Np), Imsekwa(Lm), Neorchung (Sk,Bh), Yelkung(Lp), Chuna Phala (Gn).

Description: Large sized evergreen tree. Bark light grey, blaze pink and white, streaky with a few drops of milky juice. Leaves clustered near ends of branches. Coriaceous, ovate or ovate - oblong. Blade 15 to 30 cm petiole 25-38 mm. Flower December - January, fruits May-June fleshy.

Habitat: Lower hills of Yuksom on rocky and dry slopes ascending upto 1200 m. One tree was observed at 1775 m in Yuksom lopped for fodder.

Uses: Fruits are eaten. The seeds yield white vegetable butter. This butter is eaten by villagers as vegetable oil. This is also sold in the market. The leaves offer good fodder for cattle during lean period. This is a good species to rehabilitate drier, rocky hill slopes.

(n) *Litsaca citrata.* Siltimur(Np), Warekpa(Lm), Nanghri(Sk,Bh), Tanghercher kung, Samkung (Lp), Dom (sh).

Description: Small tree of plains and hills, bark smooth, whitish brown, green on the large branches. Leaves membranous 10-18 cm long, lenceolate, fresh bright green above, dull glaucous beneath. The whole plant has delightful fragrance of oranges. Flowers December-February, Fruits March-May.

Habitat: Occurs in sub-Himalayan tract from plains to hills upto 2700m. Common in Yuksom.

Uses: The local people use the fruit (dried) for nausea and giddiness. The old people have seen wine distilled out of its fruits and used in gastritis. In the past as per local people, oil was also extracted for cooking. The fresh fruits are made into pickles.

(o) *Evodia fraxinifolia.* Khanakpa(Np), Khunu (Sk,Bh), Kunu (Lp).

Description: A moderate sized strong aromatic tree. Bark smooth light grey with horizontal lenticels. Leaves glabrous, leaflets 3-6 pair 10-22 cm long, oblong-lanceolate, acuminate, sometime slightly crenulate, base often oblique. Flowers May-June. Fruits October-December. Inflorescence stout. Young shoot, leaf rachis and fruits are red.

Habitat: In Yuksom it is found between 1200 m to 2100 m.

Uses: The fruits are used in pickles. Also used as medicine for dysentery.

(p) *Zanthoxylum acanthopodium.* Timur (Np), Midhing (Lm), Timbur (Sk, Bh), Nangryupot (Lp), Yarma (Sh).

Description: An erect shrub 1.2 to 3.0 m high with stem upto 7.5 cm diameter. Young glabrous or shortly pubescent. Stem and branches armed with conical prickles. Bark greenish ashy, smooth, with numerous pale circular raised lenticels. Blade greenish. The whole plant is very aromatic. Leaves imparipinnate, 11-25 cm long rachis with faliaceous green wing upto 5mm broad. Rachis and midrib usually armed with straight prickles. Leaflet 5-13, sub equal 3.8 to 7.5 cm long oblong, lanceolate or the basal pair often ovate acute, serrate with a bland at the base of each tooth. Main lateral nerves distinct 10-20 pairs. Flowers in June and fruits in September -November.

Habitat: Outer Himalaya upto 2250 m in Yuksom forest.

Uses: The fruit is sweetish, bitter, hot tasty and digestible appetiser and anthelmintic, removes pain tumours and abdominal troubles. Useful in eyes and ear disease, headache, heaviness, leucoderma, asthma, trouble of spleen, difficult micturation (Ayurveda). The seeds are sharp with a good taste and smell, tonic, astringent to the bowels and useful in diarrhoea, carminative, good in brain diseases and insanity, useful in stomatitis, strengthen the liver, purify the blood remove foul smell from the mouth (Yunani). Local people use small quantity for pickles. They also use it to kill leeches in the forest during rains.

(q) *Svmplocos theifolia.* Kharane (Np), Mutsang (Lm), Sembau Kung (Lp).

Description: A middle sized tree Bark brown Blaze white. Leaves glabrous, blade 7 to 15 cm. Petiole 10 mm. Nerves prominent above. Flowers October-March, fruits November -April.

Habitat: Upper hills from 1800-3000 m grows gregariously as a secondary crop. In natural forest it is scarcely found.

Uses: The profuse seeds offer good food for the birds and sloth bear. In the past, oil used to be extracted for cooking as per the people of this area.

(r) *Garcinia cowa.* Roxb Kaphal(Np).

Description: An evergreen medium sized deciduous tree. Drooping branches. Leaves broadly to elliptically lanceolate at both ends acuminate 10 to 12 cm long. Nerves faint very oblique 6-10 pairs. Flower yellow or reddish yellow. Flowering time December. Fruit May-June, size of a cherry, mesocarp reddish yellow.

Habitat: Lower and Middle hill forest found on way to Tshong. In flower during December at 1725 m.

Uses: Fruits edible. The gum-resin is used medically. Many local people are unaware of this species.

(s) *Pyrus pashia Syn* pyrus variolosa Mehel(Np), Thumbausey (Lm), Lupey (Sk, Bh), Leekung (Lp).

Description: A middle sized tree, young parts woolly (Barren branches spinescent). Leaves ovate or ovate-lanceolate, acuminate, crenulate, glabrous turning red when old. Flower March-April, fruit August-September. Fruit nearly globose when ripe 20 to 30 mm dia, crowned with base calyx.

Habitat: in Sikkim 800 m to 2400 m. Abundant around Yuksom.

Uses: The fruits are crushed, juice extracted under pressure and boiled to get an extract, locally called 'chook' which can be preserved for years and used as medicine for dysentery. The fruits are also preserved in the form of pickles.

(t) *Rhus semialata.* Bhakimlo(Np), Essebha(Lm), Tyoga sing(Sk,Bh), Tanghir Kung(Lp), Ritchi(Sh), Khyuba phala (Gn).

Description: A middle sized deciduous tree. Bark grey with numerous reddish brown lenticels. Blaze with white sticky milky juice. Branches, petioles, underside of the leaves and inflorescence clothed with short soft brownish grey pubescence. Leaves pinnate leaflet 4-6 pairs, opposite dentate, lateral sessile. Flower pale yellow green. Flowers in June - October, fruit October - December.

Habitat: Outer Himalayan Ranges grows between 900 to 1850 m very common on old cultivation and often occurring near forest boundaries quite common in Yuksom.

Uses: The small seeds are very sour and eaten to cure dysentery. The seeds are crushed and 'chook' is made out of this to be used as medicine to cure dysentery by local people.

(u) *Citrullus colocenthes*, Indrenni(Np).

Description: Large deciduous climber. The stem is about 40 mm dia. Distinct node is formed at every point from where the old branches have fallen. Distinctly striate stem. Branchlets zigzag with alternate leaf. Leaf deeply palmately lobed. Top lobe in the form of minarette. Base cordate. Size 10-18 cm long by 7-15. Fruit ripens in November - December globose 7.5 to 10 cm dia bright red in colour making itself prominent wherever it occurs.

Habitat: It is abundantly available in Yuksom. One climber was found near the Forest Rest House in Yuksom. One Gurung family has also cultivated Indernni in his field for fruit.

Uses: The red fruit when ripe becomes very soft. The inner pulp is yellow in which black seeds of the shape of pumpkin seed but double its thickness are found. These seeds are washed in water duly rubbed with ash so that the bitter taste of the pulp is not left in seed. The seeds are then fried in pan or karai and eaten. It is oily and very tasteful. It is also said to be of medicinal value requiring investigation.

(v) *Pyrularia edulis*, Amphi(Np), Toktor kung (Lp).

Description: Small or middle sized tree. Bark thin greenish grey. Leaf buds with white silky scales. Leaves elongate elliptic. Blade 12 to 18 cm with acute base. Glabrous Flowers March - May, Fruit July - October.

Habitat: This tree can grow between 600-1800 mm. One tree is on way to Dubdi Monastery from Yuksom at 1750 m.

Uses: The fruit is pear shaped tasting like guava. The globose inner stone has wax in its kernel which bums like candle. Old people of Yuksom had the experience of using this for lighting purpose in their house in the past.

(w) *Rubus ellipticus*. Aselu(Np), Tumse(Lm), Chilum(Sk,Bh), Khaseemjeeo(Lp), Nangyan(Sh), Paula(Gn).

Description: A large shrub with long trailing angular branches. The branches and petioles are covered with dense brown bristles without glands. Stout hooked prickles on branches petioles and midrib. Leaflets three sometimes only one, almost coriaceous, broadly elliptic or ovate, obtuse or shortly acuminate, densely clothed beneath with soft grey pubescence. Flowers white in February - March, fruit in April-May yellow.

Habitat: In wasteland open forest from l000 - 2200 m in Yuksom.

Uses: The fruits are edible. Can be made into jam. They are also good food for birds.

(x) *Rubus niveusb*. Kalo Ghyampey Asilo(Np), Patangwa(Lm), Rungkong Jeeo(Lp), Theba Panla(Gn).

Description: A straggling thorny shrub with purplish stem. Silky mostly grey pubescence on branchlets, leaves and inflorescence small conical prickles on branches and petioles. Leaflets 1 or 2 pairs, membranenous deeply incised - serrate often slightly lobed generally ovate acuminate. Petal pink. Flower May - July, fruit June - October. Oblong red edible.

Habitat: Available in waste land and openings of forest in Yuksom.

Uses: Berries are edible.

Buds. Young Shoots and Leaves

(a) *Heptapleurum impressum*. Chinde(Np), Sang(Lm), Jaysing(Sk,Bh), Sang Jam Kung(Lp), Khaareydhum(Gn).

Description: Large tree. Leaves digitable. Leaflets lanceolate entire, those on young trees often distantly serrate, upper side shining, the nerves impressed, underside softly stellate - tomentose. Blade 15-25 cm. Common petiole is 12-40 cm. Flower May - August. Fruits September-December.

Habitat: In Yuksom 1800 to 3000 m. A lopped tree was observed on way to Dubdi at 1750 m.

Uses: Young shoots make good vegetable and pickles. Leaves are also lopped for fodder.

(b) *Bauhinia vareigata.* Koirala (Np), Chirabey (Lm), Rhar Kung (Lp), Reysing (Sk, Bh) Kroidhum(Gn).

Description: A deciduous middle sized tree. Bark grey with numerous grey lenticels. Blaze pink and white streaks. Leaves as long as broad 11-15 nerved. Flower February -April, white except for a purple patch on the standard petal, when tree is leafless.

Habitat: In drier lower hills upto a height of 600 m in Yuksom.

Uses: For local packing purposes. Young leaves and flowers-buds are quite in demand locally for their culinary values.

(c) *Ficus infectoria*. Kabra(Np).

Description: A large widely spreading deciduous fast growing tree with long young shoots that appears in February-March. The young shoots are covered with equally long membranous stipule. It is sometimes epiphytic. Bark light grey and shiny. Leaves thinly coriaceous, glabrous, shining ovate-oblong shortly acuminate, margin undulate base acute, blade 7.5 to 15 cm petiole 2.5-7.5 cm long.

Habitat: It is mostly cultivated in and around Yuksom.

Uses: The shoots are made into pickle. It is also lightly boiled and dried and preserved to be added as souring agent to curry. The leaves form a very good fodder for cattle especially during lean period.

(d) *Laportia terminalts*. Patle sisnu (Np), Sasungey (Lm), Sochya (Sk,Bh), Kunlobee (Lp), Sanyuma (Sh), Pocho Bhoarey (Gn).

Description: A shrub upto 60 cm high armed with stinging hairs specially on leaves and young shoots.

Habitat: Very common in Yuksom.

Uses: The young leaves and shoots that come up in March-April are boiled and eaten as dal. Preferred for blood pressure patients. This is collected almost the whole year in Yuksom for vegetable.

(e) *Girardinia Palmata.* Bhangre Sisnu (Np), Hyagma (Lm), Dongey Sochya (Sk,Bh), Kaztobee (Lp), Lengochema (Sh), Pocho renga (Gn).

Description: A tall coarse herb armed with long stout stinging hairs, leaves large broad cordate variously lobed, often palmately coarsely toothed. Stipules large. Flowers in July in higher altitude and August in lower altitude.

Habitat: Grows densely in open areas and in moist places, as undergrowth. Occurs between 1000 to 2500 m in Yuksom.

Uses : The young shoots and flowers are boiled and prepared as soups. Preferred for blood pressure patient. The bark gives good fibre for making ropes, twine etc, and for coarse cloth like gunny for making bags. The leaves and shoots are also boiled and fed to cattle and pigs.

(f) *Urtica spp.* Gharia Sisnu(Np), Pangyokpa(Lm), Milleryappa(Sk,Bh), Kaiyang(Lp), Sekcheme(Sh), Dhir pocho(Gn).

Description : Dense shrubs, monoecius or dioecious pubescent, stinging hairs copious. Leaves ovate-cordate oblong or lanceolate serrate usually free.

Habitat : From 1000m to 2500m. Comes densely in open areas. Some villagers have cultivated in their field.

Uses : The lower and young shoots are boiled and eaten by high blood pressure patients. The roots are ground and mixed with quartz stone dust, harchur (Viscum articulatum) and root of chipley (Pouzolsia sp) and plastered to cure fractured bones. Leaves also boiled and fed to pigs.

Decorative

(a) *Dry mushroom*. Sukha Chew(Np).

In natural forest brace like formation of various shape, sizes and colour in dry condition are found either on dry branches or dead wood as natural show pieces or are painted and produced as attractive decoratives. They belong to poolyporacea or thyloporaceac group of fungi. These are not collected in Yuksom.

(b) *Anaphalis triplinervis*. Bhukiphul(Np), Khawa Rang(Lm), Reep(Lp).

Description: Stem rather stout often flexuous leafy and leaves beneath densely clothed with while wool. Leaves 7.5 to 20 cm ovate or elloptic oblong acute amplexicaul 3-5 nerved cobwebby above. Head in 12 mm to 20 mm dia in open or close corymb.

Habitat: These are found in open places in Yuksom.

Uses: The flowers are collected dried and used as pillow stuff in villages. The dried flowers are used as decoratives in natural condition which last for about 2 years without deterioration. This is however, not collected by locals.

(c) *Anaphalis contoria*. Buki Phul(Np).

Description: Stem 15 to 60 cm stout or slender branches prostrate or ascending leafy. Leaves both surface cottony 12 to 24 mm very crowded. Narrowly linear or oblong and nerved. Head 5 mm dia. Subglobose in dense simple or lobed, contracted corymbose clusters. White or yellowish.

Habitat: Available in the open areas of forest and outside in Yuksom.

Uses: Flowers in dried condition collected and used as pillow stuffs.

(d) *Pollinia mollis* (Erianthus). Memkesh(Np), Yang(Sk,Bh), Parphek(Lp).

Description: Stem 30-90 cm slender, tufted leafy chiefly at the base. Leaves narrowly lanceolate, sheath glabrous or ciliate. Spikes, 3-10, 2.5 to 5 cm long crowded, flexuous, joints and pedicels slender beautifully silvery silky with very long hairs. Spikelets pale glume thin villous hairs much longer than itself lower half dark brown upper pale. Hooker described it as almost intermediate between pollinia and Erianthus.

Habitat: Temperate Himalaya - Sikkim 1550 m to 2450 m. In Yuksom these are seen on rocky road cuttings, exposed rocks.

Uses: The spikes in small bundle made beautifully bouquets of dry flowers that last for more than a year as decorative. But people here do not collect.

(e) *Moss*. Jhau(Np), Tamsung/Yangbew(Lm), Mamah(Sk,Bh), Amal-Sungphon(Lp), Jhyambo(Gn).

Description: These are found as secondary growth attached on trees and on rocks in the forest of Yuksom.

Uses: Moss is used mostly in nursery as moisture absorbent. Also used as decorative.

(f) *Lycopodium clavatum*. Nagbelli(Np), Ghisingdermu(Lp), Sisingderma(Sk,Bh).

Description: A trailing creeper growing in open areas on ground or rocks spreading with roots at every node. Stem green covered with dense thick spiny leaves all over except on the under surface. Spikes on the tips of each branchlets on maturity the spikes burst the powder like spores.

Habitat: Growing on open areas between 1850 m to 2750 m.

Uses: The entire plant is collected for use as decoratives. The spores can be used for various purposes including powder.

(g) *Kalamay Uneu*. Fern: Washeing(Lm), Chya Kim(Sk,Bh), Zeethung Krok/Purmulasong(Lp).

Description: These ferns grow in dense clumps with spreading branches and have 1-1.5m long woody stems.

Habitat: It grows in the hills from 1650 m - 2450 m in more or less open areas. A clump is seen near Dubdi Monastery.

Uses: In the past writing pens were out made out of stem. The stem can also be used as sticks in making 'Stools (*Muras*)'. The young shoots are collected in huge quantity from Darjeeling hills as decoratives. Here, this fern is not much in use.

Mushrooms

There are 106 varieties of edible mushrooms in India. Out of them about 11 edible varieties were seen in Yuksom. They not only form delicacies in food of local people, but are also a good source of nutrition. The following varieties are described here:-

(a) *Agaricus silvaticus*. Kalungey Cheu(Np), Peepath(Lm), Sangey Jhyamu(Sk,Bh).

Description: With the onset of monsoon in May this particular mushroom starts appearing on ground in scattered groups. The pileus usually 3.5 to 10 cm dia convex when young becomes flat when old. Upper surface whitish or brown. The stipe (stem) is 3-12 cm long 1-1.8 cm dia. The villagers of Dosthang informed of having single variety known as ekley Kalungey which is 20-25 cm diameter and 35-50 mm dia stipe and 20-30 cm high.

Habitat: It is found in Yuksom upto 1300 m altitude.

Uses: This is available for about two months in May and June. Local people call it poor man's meat because of its taste. The ekley variety is said to appear in September and October. One such mushroom is enough for a family of 6-7 members.

(b) *Pleurotus Saior Caju*. Kanney Cheu(Np), Nak Thuk(Lm), Kayley(Sk), Tumung Dorby(Lp), Shyamu(Sh).

Description: Sporophores are found growing in dead trees either fallen or standing. Specially on Dudhilo, Nebharo, Gobre, Saur and Khanankpa, Pileus kidney shaped. Some are round with short stalk white or grey 2.5 to 10 cm dia.

Habitat: In natural forest 1500 m to 2450 m on dead wood. They are said to sprout after lightning during August -September.

Uses: It is one of the most tasty mushrooms. In the past one could find about 5 kg in one harvest.

(c) *Chipley Cheu* (Np), Warreng Lemba(Lm), Ghetka Shyamu(Sk,Bh), Phukip Dur(Lp).

Description: These mushrooms have their pileus 2 cm to 3 cm dia 2.5 to 3 cm stalk. The top surface is glistening reddish with brown and white margin. Since the mushroom is slippery, hence the name. They grow gregariously on rotten stumps or logs during rains.

Habitat: Available in the forest of Yuksom from 1500 to 2450 m altitude.

Uses: These are mostly collected as vegetable from rotten trees like sour and rani champ.

(d) *Jhari Cheu.* (Np).

Description: With the monsoon this particular species of mushroom occurs in abundance on humus in forest. Their colour is greyish white with pileus having 20 mm maximum dia.

Habitat: These occur in the hills above 1800 to 2000 m in the natural forest of Yuksom.

Uses: Collected as vegetable.

(e) *Pleuratus ostratus.* Syn Pleurtus salignus Chamrey(Np), Muya Lappa(Lm), Khomtu Shyamu (Sk,Bh), Kombu Dur(Lp).

Description: Sporophores growing on dead trunk of Tarsing or its branches. Whitish large tough when old Pelius 8 - 20 cm or more broad spathulate to kidney shaped. White grey or sometime yellowish. Surface smooth, margin incurved. Stipe 1-3 cm eccentric or laeral 0.5 to 2 cm thick.

Habitat: Found in forest of Yuksom.

Uses: Tasteful when cooked though tough to chew.

(t) *Polyporous umbellatus.* Per Exfr, Syn, Boletus kumbellatus Pers, Grifola ramosissima Scop Ex. Murr, Polypilus Pers Exfr, Thankrey(Np), Thukchuk, Shyamu(Sk,Bh).

Description: Sporophores growing at the base of tree specially phalant(Q. lineata), stipitate 7-20 cm in dia compound formed by a main stipe with numerous small branches each bearing a single centrally attached pileus arising from a sclerotiod base. Pileus whitish to smoky brown small round fleshy surface fibrillose or smooth stipe, white several stipe of the adjacent pilea united at the base to form a large clump.

Habitat: High altitude mushroom found rarely above 2500 m in the forest of Yuksom.

Uses: This is also collected as vegetable.

(g) *Hieun Cheu.* (Np). Nang Nakthuk(Lm), Khau Shyamu Sonangdur(Lp).

Description: This mushroom brown in colour grows profusely on dead woods of Asare, kaulo and other dead wood and also on ground after snowfall. Villagers are said to collect upto 50 kg in one season.

Habitat: Above 2500 m altitude after snowfall.

Uses: Very popular vegetable for people in Yuksom.

(h) *Katusey Cheu.*(Np), Singdambey(Lm), Sgey Jhamu(Sk,Bh), Kusue Dhur(Lp).

Description: This mushroom is found on dead and rotten Castanopsis tree for which the name is after this tree.

Uses: This mushroom is quite tasteful and are collected during rains for vegetables.

Diplazium (Fern Shoot, Ningro)

(a) *Diplazium polypodiades* Mett, Sauney Ningro (Np).

Description: The axis erect, stout densely clothed at the crown with long brown, fibrilous scales, 25mm long, stipes densely tufted, stout green 30 cm or more long, fronds 90-120 cm and wide, bipinnate with 8-9 primary pinnase on each side below the simple one. The lower ones generally distant. The upper surface looking coarse and pale green.

Habitat: Most common fern in almost all moist and shady places and banks of jhoras in Yuksom.

Uses: The young shoots are covered with fibrilous brown hair and they are popular vegetable during July-August (Saun). That is why it is called Saun Ningro.

(b) *Pahakhuttey* (Np).

Description: Caudex is short, leaves 0.5-1.0 m long, fronds 0.3-2.5 cm long, rachis 16 mm. Primary pinnae sub opposite to alternate upto 13 pairs, lower ones 7-18 cm. Upper ones tapering 20-25 mm. Baby plants develop like small knurled bulbs on the rachis in December. On maturity it drops on the ground and forms independent plant, normally 3 or 4 such plantlets are developed from one mother plant.

Habitat: It is found in shady and moist areas above 1750 m in Yuksom.

Uses: The shoots are brought in large quantity in the market during August-September. This is a favourite vegetable of the people of Yuksom.

(e) *Kali Ningro* (Np).

Description: Caudex is dark brown about 7.5 cm dia and 15 cm long with fibrous and long roots. The leaves look like that of Sauney but is light green in colour and soft in texture. The position of the secondary pinnae is alternate of sub-opposite as against opposite of Sauney.

Habitat: These are at altitude above 1750 m in jhoras and depressions.

Uses: The young shoots which are dark green in colour have long white streak on the side. The shoot is bit slippery.

Yams

(a) *Dioscorea hamiltonii*, D.hookerii, Ban Tarul (Np).

Description: A climber with angled glabrous stem, twining to the right. Leaves opposite on thicker stem often alternating towards the end. Bulbils abundant. Tubers long and deeply penetrating into the ground. Flesh white and edible.

Habitat: Available in the private forest along the road to Tshong village.

Uses: It is most esteemed among wild yams but it is difficult to dig. However, pits are dug upto 1.2m deep to extract the tuber.

Fibre

(a) *Edgeworthia gardnery*, Kagate or Argeli (Np), Dhenok (Lp).

Description: A large shrub. Branches long scrambling. Leaves glaucous lanceolate 10-18 cm 2.5-4 cm. Flowers in December-May in dense heads. Fruits June-September.

Habitat: Local people use its bark for making paper and ropes for carrying loads and also for tying cattle etc.

(b) *Daphne cannabina* , Kagatey or Lokhtey (Np), Dhenok (Lp).

Description: It grows upto 4 m height. Bark & shoot, rough scaly. Branches horizontal and drooping, dense. Leaves dark shining green, veins invisible on upper surface. Lower surface pale green. Flower in December - February white with pink tinge, fruits July, red in colour.

Habitat: It is available in Yuksom above 1850 m upto 2500 m.

Uses: Its bark also is used locally for ropes only. This however, has potential as raw material for paper along with Edgeworthia.

Suggestions for Qualitative Improvement of NTFPs.

(a) The potential for on-farm cultivation of NTFPs which are of economic importance to people and involve villagers also for its propagation specially those medicinal plants that are getting rare and extinct be exploited.

(b) Collection and sale should be encouraged for cultivated items and those allowed from wild on sustainable basis only.

(c) Research needs to be conducted to study the various uses of NTFP not covered so far in other areas.

(d) NTFP like 'tama' (Shoot of bamboo) mushroom, ningro (Diplazium), Sisnu (nettles) and pickle could be served in local hotels as delicacies to popularise these products. Items like pickle can be sold also with local brand which can become a source of income for the local people.

(e) NTFPs are resources to sustain local needs of life and it can also act as source of economic benefit to the local community. It has great potential but its sustainable harvest strategy has to be developed. Therefore, community involved management and monitoring of NTFP could be one approach followed at organisational levels.

References

1. Margalef, Ramon; *Perspectives in Ecological Theory*, University of Chicago Press, Chicago, USA, 1968(PA).
2. Bennett, John. W; *The Ecological Transition*, A. Wheaton & Co. Exeter, Great Britain, 1976 (P.25).
3. Davis, E. Wade; From *Ethnobotany: Evolution of a Discipline*, by Schultes and Reis(ed) Diosconrides Press, Oregon, USA, 1995 (P.40).
4. Alcorn, Janis.B; From *Ethnobotany: Evolution of a Discipline*, by Schultes and Reis (ed), Dioscorides Press, Oregon, USA, 1995. (P.23).
5. Ford, R.I.; *Ethnobotany: Historical Diversity and Synthesis,* Anthropological papers, No 67, Michigan Museum of Anthropology, 1978. (P.33-49).
6. Ricour, P.; *The Model of the Text :Social Research*: 5, University of Chicago Press, Chicago, 1971, (P.529).
7. Posey, D.A; *Indigenous Ecological Knowledge and Development*, CO West View Press, USA, 1983, (P.226).
8. Alcorn, J.B. *Huastec Mayan Ethnobotany*, Austen, University of Texas Press, 1984.
9. Speight, M.R. *The Potential of Ecosystem Management*, in 'Agriculture' Ecosystems and Environment; 10, 1983. (P.183-199).
10. Lynch, O., and Alcorn J.B. *Empowering Local Forest Managers*, World Resources Institute, Washington, DC, 1991.
11. Wilson, G.F., and B.T.Kang; Developing stable and productive biological cropping. In *Biological Husbandry*, (Ed). B. Stonehouse. Boston: Butterworths (1981).
12. Dimbleby, G.W.; *Plants and Archaeology*, Atlantic Highlands, Humanities Press, New Jersey, USA, 1978.
13. Boster, J.S.; *Ethnobotany in Neotropics*, VoL.I., Bronx, NY: New York, 1984 (pp. 34).

14. Barlett, P.F., ed. *Agricultural Decision Making: Anthropological Contributions to Rural Development*, Academic Press: New York, 1980.
15. International Potato Center, *Breaking New Ground: Anthropology in Agricultural Research*, Lima, Peru, 1984.
16. Schultes, R.E., and S von Reis, ed. *Ethnobotany : Evolution of a Discipline,* Dioscorides Press, Portland, Oregon, USA, 1995.
17. Naranjo, P., *Medicinal Plants in Ecuador*, Ecuat de Medicina, Quito, Ecuador, 1984.
18. Swain, T. ed., *Plants in the Development of Modern Medicine*, Cambridge, MA: Harvard University Press, 1972.
19. Farnsworth, N., and A. Bingel. *New Natural Products and Plant Drugs*, Springervelag Press, New York.
20. Von Reis and Altschul, S., *Drugs and Foods from Little Known Plants*, Cambridge, MA: Harvard University Press, 1973.
21. Balandran, M.F., *Natural Plant Chemicals: Sources of Industrial and Medicinal.*
22. Duke, J., *Medicinal Plants*, Science Magazine 229: 1038 ; USA, 1985.
23. Klayman, D.L., Qinghaousu: *An Antimalarial Drug from China*, Science Magazine 228: 1049; USA, 1995.
24. Godbole, A. Ajay Rastogi and Peishengji, *Applied Ethnobotany in Natural Resource*. ICIMOD Papers, 1996.
25. Padoch, C. and Jong, W.de, *The Home Gardens of Santa Rosa.*, Economic Botany, 42(2) : 166-175, 1991.
26. Ninej, Y.K., *House Hold Gardens*, International Potato Center, Lima, Peru, 1984.
27. Saklani, A. and S.K. Jain, *Cross cultural Ethnobotany of NE India*, Deep Publishers, New Delhi, 1994.
28. Brownrigg, L., *Home Gardening in International Development*, The League for International Food Education, Washington, D.C., 1985.
29. Fernandes, E.C.M. and P.K.R Nair, *An Evaluation of the structure and Function of Tropical Home Gardens*, Agricultural Systems, 21: 279-310,1986.
30. Subba, J.R, *Agriculture in the Hills of Sikkim*, Sikkim Science Society, Gangtok, 1984.
31. Barran, J., *Subsistence Agriculture in Polynesia and Micronesia.*, Bishop Museum Bulletin No 223, 1961.
32. Michon, G., *Village-forest Gardens in West Java*, International Centre for Research in Agroforestry, Nairobi, Kenya, 1983.
33. Altieri, M.A. and J.Farrell, *Traditional Farming systems of South-Central Chile*, Agroforestry Systems, 2 : 3-18, Nairobi, Kenya, 1984.
34. Fernandes, E.C.M.; A. *Oktingati and J. Maghembe*, The Chhaga Home Gardens, Agroforestry Systems 2 : 73-86, Nairobi, Kenya, 1984.
35. Okafor, I.C., and E.C.M. Fernandes, *The Compound Farms in SE Nigeria*, Agroforestry Systems, 5 : 153-168, Nairobi, Kenya, 1987.

36. Oduol, P.A. and J.R.W. Aluma, *The Banana-Coffee Robusta : Traditional Forestry System in Uganda*, Agroforestry Systems, 11 : 213-216, Nairobi, Kenya, 1990.
37. Millat-e-Mustafa, *The Ecology and Management of Traditional Home Gardens of Bangladesh*, Ph.D. Thesis. University of Wales, Bangor, UK, 1996.
38. Harmer, K., *The Potential of Mixed Orchards in Britain*, Undergraduate Thesis, University of Wales, Bangor, UK, 1991.
39. Tarnutzer, A. and C. Bretting, (ed), *Participatory Rural Appraisal on Rural Livelihood Systems in Sikkim*, Gangtok and Zurich, 1994.
40. Paljor, S., Feed and Fodder Resources of Sikkim, in, *Sikkim : Perspectives for Planning and Development*, by S.C. Rai, R.C. Sundriyal and Eklabya Sharma, (eds), Sikkim Science Society, Tadong, Gangtok, 1998.
41. Mergen, F., *Research opportunities to Improve the Production of Home Gardens*, Agroforestry Systems, 5 : 57-67, Nairobi, Kenya, 1987.
42. Soemarwoto, O., Home Gardens : A Traditional Agroforestry System with a Promising Future. In, *Agroforestry : A Decade of Development*, by Steepler, H.A. and P.K.R.Nair, (eds), International Centre for Research in Agroforestry, Nairobi, Kenya, 1987.
43. FAO, *Household Food Security and Forestry* (An Analysis of Socio-economic Issues), FAO, Rome, 1989.
44. Brownrigg, L., *Home Gardening in International Development : What the Literature Shows*; The League for International Food Education, Washington, D.C., USA, 1985.
45. Lama, Mahendra, P., (ed) *Sikkim : Society, Polity, Economy, Environment*, Indus Publishing Company, New Delhi, 1994.
46. GBPIHED, *Annual Report* 1996-97, Ministry of Environment & Forests, Almora, 1997.
47. Messer, Ellen., Methods for Studying Determinants of Food Intake, . In *Research Methods in Nutritional Anthropology*, by, Gretel. H. Pelto ; Pertti. J. Pelto and Ellen Messer; The United Nations University, Tokyo, 1989.
48. Tamang, J.P., P.K. Sarkar, and Hesseltine, C.W., Traditional Fermented Foods and Beverages of Darjeeling and Sikkim - a review, in *J. Science of Food and Agriculture* 44 : 375-385. 1988.
49. Tamang, J.P., Indigenous Fermented foods of Sikkim Himalaya : Socio-Economical prospective. In *Sikkim-perspectives for Planning and Development* (Eds : S.C. Rai, R.C.Sundriyal and E. Sharma), pp 513-522, Sikkim Science Society, Gangtok, 1998.
50. Kwon, T.W., The Role of Fermentation Technology for the World Food supply. In: *Lactic Acid Fermentation of Non-Dairy Food and Beverages*, (eds) G.H. Loe, J. Adler Nissen and G. Barwald. P. 1-7. Ham Lim Won Publication, Seoul, Korea, 1994.
51. Steinkraus, K.H., *Nutritional Significance of Fermented Foods*. Food Research International 27: 259-267, 1994.

52. Camphell-Platt, G., *Fermented Foods - A World perspective*. Food Research International 27 : 253-257, 1994.
53. Tamang, J.P., Role of Microorganisms in Traditional Food Fermentation Technology in the Himalayan Regions. *Proceeding Abstract of the Tenth International Conference on Global Impacts of Applied Microbiology & Biotechnology* ; Elsinor, Denmark. P.61, 1995 b.
54. Tamang, J.P., Fermented Soybean Products in India. In : *Proceeding in the Second International Soybean Processing and Utilization Conference, Bangkok* (ed) A. Bucharan, P.189-194. IFRPD, Kasetsart University, Thailand, 1996.
55. Tamang, J.P., Studies in the *Microflora of Some Traditional Fermented Foods of Darjeeling Hills and Sikkim*, Ph.D. Thesis, NBU, India, P.224, 1993.
56. Sarkar, P.K., Tamang, J.P., Cook, P.E. & Owens, J.D., Kinema - *A Traditional Soybean Fermented Food : Proximate Composition and Microflora*, Food Microbiology 11: 47-55, 1994.
57. Sarkar, P.K., and J.P. Tamang, *The Influence of Process Variables and Inoculum Composition on the Sensory Quality of Kinema*. Food Microbiology, 11: 317-325, 1994.
58. Tamang, J.P. and Nikkuni, S., *Selection of Starter Culture for Production of Kinema*, A Fermented Soybean Food of the Himalaya. World J. Microbiology and Biotechnology, 12: 629-635, 1996.
59. Tamang, J.P., Kinema - A non-salted Fermented Soybean Food of Darjeeling Hills and Sikkim. *Proceeding abstract of Third Asian Symposium on Non-Salted Soybean Fermentation*, Akita, Japan, 14, 1994.
60. Sarkar, P.K., and Tamang J.P., Changes in the Microbial and Proximate Composition during Natural and Controlled Fermentations of Soybeans to Produce Kinema. *Food Microbiology. 12 : 317-325*, 1995.
61. Tamang, J.P., *Study of Traditional Fermented Foods Production in Darjeeling Hills and Sikkim, with Emphasis on Kinema*. Post Doctorate Dissertation, United Nations University, Tokyo, Japan, 1995.
62. Nikkuni, S., Karki , T.B., Vilkhu, K.S., Suzuki, T., Shindoh, K., Suzuki, C. and Okada, N., Mineral and Amino Acid Contents of Kinema, *A fermented Soybean Food prepared in Nepal. Food Science Technology International 1 (2) : 107-111*, 1995.
63. Tamang, J.P., 1995, *ibid.*
64. Karki, T., Okada, S., Baba, T., Itoh, H., and Kozaki, H., Studies on the Microflora of Nepalese pickles Gundruk. *Nippon Shokuhin Kogyo Gakkaishi*, 30 : 357-367., 1983.
65. Tamang, J.P. and Sarkar, P.K., Sinki : A Traditional Lactic Acid fermented Radish tap root product. *J General & Applied Microbiology, 39 : 395-408*, 1993.
66. Karki, T., Food Processing Industries in Nepal. In : *Proceeding of International Workshop on Application and Control of Micro organisms*

in Asia, Tokyo, March 1994, Science and Technology Agency, Japan, 1994.

67. Tamang, J.P. and Sarkar, P.K., Microbiology of Mesu a traditional fermented bamboo - shoot product. *International J Food Microbiology, 29: 49-58*, 1996.
68. ________________________., Microflora of Murcha an amololytic fermentation starter, *Microbios*, 81 : 115-122, 1995.
69. Tamang, J.P., Thapa, S., Tamang, N. and Rai, B., Indigenous fermented food beverages of Darjeeling Hills and Sikkim : Process and Product characterization. *J Hill Research*, 9 : 401-411, 1996.
70. Tamang, J.P. and Sarkar, P.K., 1995, *ibid.*
71. Tamang J.P. it al, 1996, *ibid.*
72. Tamang J.P., and Sarkar, P.K., 1995, *ibid.*
73. Yonjan, H. and Tamang, J.P., Consumption Pattern of Traditional Fermented Foods in Sikkim Himalaya, Journal *of Hill Research 11 (1) : 112-115*, 1998.
74. Tamang, J.P. and Tamang, N., *Traditional Food Recipes of the Sikkim Himalaya.* A Report, Sikkim Biodiversity Ecotourism Project, Tadong, Sikkim, 1998.
75. Hoffman, Albert., Medicinal Chemistry's Debt to Ethnobotany. In, *Ethnobotany : Evolution of a Discipline*, (eds) R.E. Schultes and S. Reis, Dioscorides Press, USA, 1995.
76. Purohit, A.N., Preface to book : *Medicinal Plants of the Sikkim Himalaya,* GBPlHED, Bishan Singh Mahendrapal Singh, Dehradun, 1994.
77. Rai, L. and Sharma, E., *Medicinal Plants of Sikkim Himalayas*, GBPIHED, Bishan Singh and Mahendrapal Singh, Dehradun, 1998.
78. *ibid.*
79. *ibid.*
80. *ibid.*
81. IUCN, WHO and WWF., *Guidelines on the Conservation of Medicinal plants*, IUCN, Glanal, Switzerland. 1993.
82. *ibid.*

6

Upper Rangit Basin : Human Ecology of Eco-Tourism

SECTION I
INTRODUCTION AND CONCEPTUAL ASPECTS

Mental attitudes differ from person to person, so do human wishes. Motivation for travel will also differ from individual to individual. Nevertheless, an attempt can always be made to classify human motives for travel :-

(a) **Rest and Relaxation**. When the humans work in a society of tension we want to get away from it, and seek physical and mental relaxation elsewhere, though temporarily.

(b) **Educational and cultural**. Humans wish to see the way of life of other people, to study the beautiful natural sights of other countries, to acquire better knowledge and background of people living there, to visit cultural, scientific, technical and spiritual institutions of learning in other places, and to appreciate art, music, literature and cultural heritage of other societies.

(c) **Social and historical**. Some are keen to visit historical sights in order to participate in history and better appreciate historical events, to gain new experiences by meeting new people, and to create personal esteem.

(d) **Health and Sports**. Another group may like to visit health centres and mountain resorts, to enjoy salubrious weather, to take part in sports or to join mountaineering expeditions.

(e) **Recreational Activities**. Activities such as swimming, trekking or hiking may be the pull factors for many.

Depending on choice and economical factors, tourism is sub divided into three categories - domestic, regional and international. Both, in terms of volume and also in terms of receipts, domestic tourism far exceeds the other two types. Regional tourism is much higher than international tourism in the case of both the variables. (Mridula and Datt[1], 1991).

While domestic tourism involves redistribution of incomes within a country, international tourism adds to the resources of the host country.

For a number of developing countries, their natural environments continue to be source of significant economic benefits, attracting international and domestic visitors often in search of an authentic natural and some times a cultural experience. Detailed information on the number of visitors visiting natural areas and the amount of resultant economic activity in the developing countries, is both, difficult to obtain. Tourism associated with natural and protected areas has been, and continues to be growing sector in the global tourism industry.(Whelam[2], 1991; Brandon[3], 1996). Of the estimated 55 bn Dollars in tourism receipts in the Third World, a significant portion of this in the early 1980s was the result of nature tourism activities (Whelam[4], 1991).

Although not always concerned with the protected areas, it is estimated that annual global expenditure on adventure-related travel amounts to 110 bn Dollars (Adventure Travel Society[5], 1998). Conservative estimates of the growth in demand for nature-related tourism range from 10-15 percent, while more optimistic forecasts go up to 30 percent (Brandon[6], 1996), the variation is due to primarily the differences in the definition of the types of tourism included. Although difficult to generalise, for many developing countries, their natural heritage is a primary attraction for international visitors and the increasing numbers of domestic tourists.

Eco-Tourism

Associated with the growth of nature, related travel sector is the facet of eco- tourism. The Ecotourism Society defines ecotourism as, "Responsible travel that conserves natural environment and sustains the well being of people." (Lindbergh and Hawkins[7], 1993). In other words, ecotourism is the symbiotic relationship between tourism and environment which benefits the socio-economic condition of the local inhabitants and at the same time promotes tourism and protects the environment.

By harbouring environmental awareness, both, among local people and visitors and by linking tourism to social and economic development, it is potentially possible to conserve the natural and cultural heritage of an area, and to improve the living standards of the host region inhabitants. Ecotourism, if developed responsibly, and managed properly, can become a powerful instrument and strategy for the biodiversity conservation. It is a form of tourism that promotes tourism-environment linkages, and at the same time, the host regions can take pride in what they have to offer for their people and the tourists.

Eco-Tourism in Sikkim

Most of the ecotourism activities in Sikkim are either in protected areas or cultural sites. Protected areas are the State Properties where ownership, control and management of resources and claims rest with the government. Rights for tourism purpose are with a variety of users including local entrepreneurs, communities commercial tour operators (residents and also from outside the area and sometimes region) and government agencies such as the Forest and Tourism Departments. Permits are often issued by the Tourism and Forest Departments of the State Government with the intention of generating revenue and regulating visitor flow. Cultural sites of tourism interest such as monasteries and their sacred forests are under the control of local institutions. There is a unique example of Khecheopalri sacred lake where the legal rights lie with the State Government but the operational rights seem to be with a local institution, viz, Khecheopalri holy Lake Welfare Committee.

The Eastern Himalaya is recognized global site of high biodiversity value, and is facing growing threats from mass tourism. Sikkim is growing in popularity, and likely to attract increasing numbers of tourists interested in the cultural sites and Protected areas of high biodiversity value. Tourism is a highly seasonal activity in the State-October through November and March through June being the two peak seasons. Since income and employment are dependent on the tourist flow, the income earned during the tourist season has to compensate for the lean tourist seasons. The seasonality factor, therefore, makes it more important to link tourism to local production systems.

As concern grows over the loss of natural heritage in the area, it appears this is an ideal time to facilitate ecologically and socially responsible tourism at key destinations to conserve the biological diversity of the area. Therefore a collaborative project of the Mountain Institute and G.B. Pant Institute of Himalayan Environment and Development with other collaborators, namely the Travel Agents

Association of Sikkim (TAAS), the Green Circle and local NGOs with communities is being implemented in the West District of Sikkim. The Government of Sikkim has also played a key role at the various stages of the project implementation. The activities and roles of each stakeholder have been presented in Table 6.1. At the heart of the project are participatory approaches that link enterprise operation with conservation action, while merging traditional cultural practices.

Human Ecological Interactions

In Sikkim the target areas are the local inhabitants and sites to be visited. The interacting groups are : the local hosts, the interacting operators, NGOs and scientific institutes - each one of them functioning within the frame work of policy laid down by the government agencies, and policy planners.

At any one point of time, all individual groups may be interacting with one or more, and some times all the agencies named above. So as to ensure the tourists' satisfaction, locals' economical enhancement and environmental preservation, it is necessary that the links between all units function unhindered. The interaction between all agencies are best explained through the Tourism Interactional Web at Figure 6.1. Such an interaction within the parameters and principles of human ecology will benefit all viz the soil and geological factors, water bodies, the mountain tops, the faunal and floral diversity, the local populace, tour operators, researchers, government agencies and the tourists themselves.

Human Ecology Interactional Web (Fig 6.1; page 246) clearly amplifies that an activity at one place (or inactivity at one place) has a profound affect on all other actors within the web. A broken link in the web may commence a chain of instability, resulting in degeneration of environmental factors on the left side of web, disorganization of agencies on the right hand side of the web, tourist dissatisfaction who interact with both, the left hand side environmental elements and right hand side human agencies, finally culminating into a human, natural and economic disaster for the local inhabitants who interact with all elements at all times.

As and when, and if at all, the indigenous ecological knowledge of the local populace is given the weightage it deserves, then the dotted lines of local populace feed back may be converted into direct undotted ones. This could be a major step towards sustainability of ecosystems themselves, and the interrelationship between each of the elements as well.

TABLE 6.1
Role, Responsibilities and Activities of Different Groups in Sikkim and Biodiversity and Ecotourism Project

(Sharma et al[9], 2000)

Activities	*Tour Operators*	*Community/NGOs*	*Government*	*TMI*	*GBIHED*
Role	To cover issues of conservation, policy and infrastructure development and management of Tourism industry as TAAS/PTDA	Preservation of natural and cultural heritage. Local stakeholders use natural resources	Government of Sikkim is responsible for policy formulation and resource control and management	Dedicated to conserve mountain environment and cultural heritage through education and development working with the community. To develop Participatory and Scientific monitoring	Focal institute to advance scientific knowledge to evolve and demonstrate integrated strategies for resource and sustainable development and to pursue applied research.
Activities/Responsibilities					
Product and Site Enhancement	Fuel-wood reduction trekking hut, garbage, litter and sanitation management,	Community forest management, conservation education and village beautification	Infrastructure development and promotion of health and safety measures.	Enterprise destination enhancement	Technical input in choice of species and planting techniques.

Activities	Tour Operators	Community/NGOs	Government	TMI	GBPIHED
	health and safety measures.				
Enterprenuerial skills and product development	Trekking staff training, new product development	Study tours for guides, lodge management, promotion of local food, micro-enterprise development	Tourism Department participation in training	Training input and organization	Green house vegetable production, composition and indigenous food promotion
Ecotourism marketing	Market trend survey to design and implement marketing strategies	Help in implementation of tourism promotion activities	Quality tourism development	Provide international experts	Provide local experts study towns and workshop.
Planning, monitoring and applied research	Policy review, business survey and participatory monitoring.	Community resource management	Policy review implementation	Policy workshop	Participatory, biological socio-economic and enterprise monitoring.
Biodiversity Preservation.	Threatened habitats and areas are not exposed to tourists.	Avoid disturbance and negative impact to wildlife.	Signboard about vulnerable areas and species.	Plan activities with no negative impact.	Carry out impact studies and monitoring.

Activities	*Tour Operators*	*Community/NGOs*	*Government*	*TMI*	*GBPIHED*
Cultural preservation	Trekking staff to inform visitors about culture and value	To be aware and feel pride about culture and tradition	Cultural site planning for tourism.	Support in managing cultural site and museums.	Support in the management of cultural sites.

TAAS - Travel Agents Association of Sikkim.
PTDA- Pelling Tourism Development Association.
KCC- Khangdzonga Conservation Committee.
TMI- The Mountain Institute.
GBPIHED - G.B. Plant Institute of Himalayan Environment and Development.

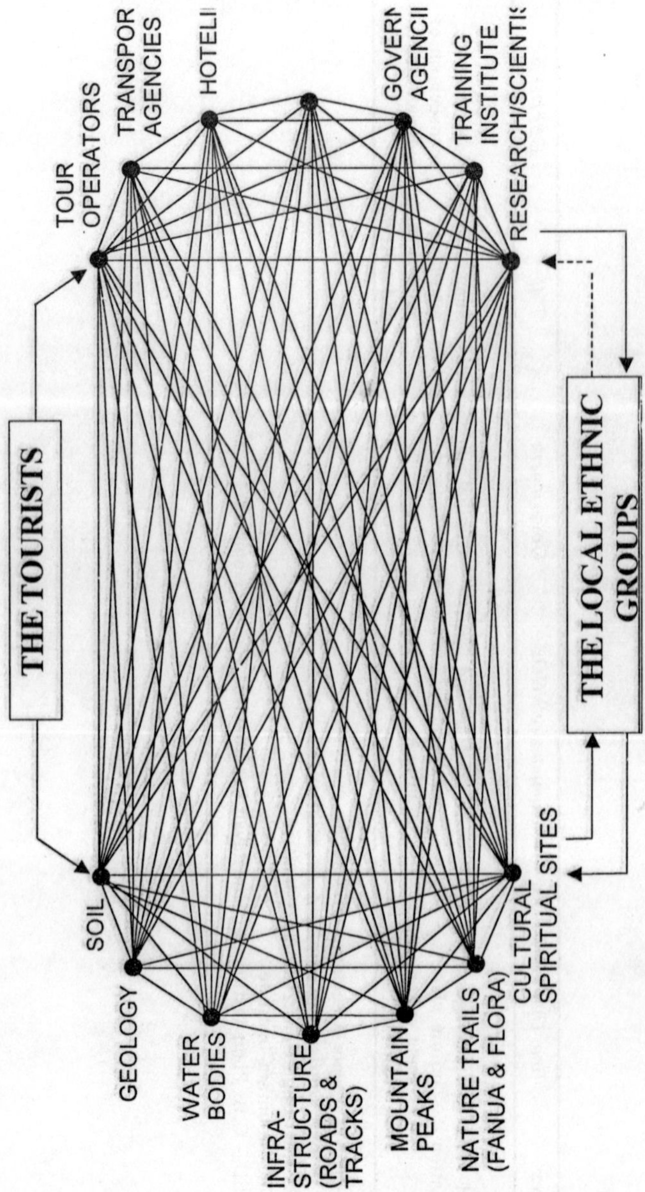

Fig. 6.1

Human Ecology of Eco-Tourism in Sikkim Interactional Web

SECTION II
TOURISM ASSETS OF UPPER RANGIT BASIN

Introduction

Lack of any form of modernization in itself is a unique attribute of the area, and when blended with natural setting, it makes Upper Rangit Basin very unique. Upper Rangit Basin constitutes areas that include heights from sub-Himalayan Regions, to the Kanchendzonga Massif. The people practice traditional customs of their respective ethnic groups and the unified Sikkimese hospitality. Its remoteness and its being on an independent salient from the State capital, adds to its attraction. The sheer beauty of the cold natural wilderness of Dzongri-Tsoka heights constituting Sikkimese polity and Tibetan culture; the rhododendron trails on the Singalila Range; athe historical assets at Yuksom and the ruins of Rabdentse; the spiritual centres at Pemayangtse, Dubdi, Tashiding and Ralang; the Upper Rangit Basin attractions may be classified into two categories, namely, natural and man-made.

Natural assets include the endowments of the natural while man-made assets are both, tangible (monasteries), as well as intangible (culture).

NATURAL ASSETS
KHANGCHENDZONGA NATIONAL PARK

History

The Khangchendzonga National Park, Sikkim may be aptly called the haven of prized musk deer, elusive snow leopard and the shy Himalayan tahr, the Shapi of Sikkim. The National Park occupies a place at the apex amongst the high altitude National Parks in the country and is endowed with one of the most magnificent high altitude ecosystems in the world. The Park which has an elevation of 1829m to over 8580 m above mean sea level was commissioned on 26th August 1977 on the basis of the floral, faunal, ecological, geomorphological importance and the wildlife potentiality of the area. (Forest Department, Government of Sikkim[15]- 1998).

Location

"This National Park lies within the latitude 27 degree 25' and 27 degree 55' North and longitude 88 degree 40' East. It is situated in the

North District where it occupies major portion of areas and in the West Districts of the State. It is a vast area extending from the portions of cold desert of Lhonak valley, ridges of Lachen in the North Districts to the historical place at Yuksom and extending upto as far as Boktok, Dhaphey Bhir and Nepal border in West Sikkim. The Western boundary of the Park runs along the international boundary between the Indian state of Sikkim and Tibet in the North and Nepal in the West Districts of the State.

Area

Originally it encompassed an area of 850 sq. kms and in the year 1996 it was expanded to 1784 sq.kms. The necessity of its expansion was essential in view of its ecological, faunal, floral, geomorphological and zoological association and importance, and to bring potential areas under the Park for providing an insurance in their protection. With the expansion in its area, the Park alone now occupies as much as 25.14 percent of the land area of the State against 11.97 percent in the past. It is proposed to be Biosphere Reserve by including additional areas in North and also the Mainam Wildlife Sanctuary.

Land Tenure

The vast expanse of the Khangchendzonga is within the reserve forests. Except a small Tibetan village of Tibetan communities settled since 1959 in a place called Tsoka which comprises 10 houses now with a total population of 90 (Contact report- September, 1998). There is no other village is settlement inside the National Park. Since this small village occupying as much as 13 acres of land only and situated in an isolated pocket within the areas under proposal for tourism zone, this will have to be considered as part of the ecosystems of the Park. The only one decimating factor taking place is the grazing by a few cattle herds in the alpine land in the Southern extremity of the Park. They were enjoying this grazing right since centuries but the impact of grazing as per the recent survey has not been so degrading to the alpine environment. This could probably be due to less number of cattle. Even though, efforts are being made to evict them to areas available outside the Park. There is also a statutory order of the State Government banning the grazing inside the Reserve Forests.

Flora

The floristic wealth is rich and diverse, both in composition and value. The forests represent variegated plant communities which include diverse vegetational types corresponding to variation of climatic and edaphic factors. The area broadly comes under Champion's and Seth's classification type SUB GROUP 11B-NORTHERN MONTANE TEMPERATE FORESTS AND GROUP 12 HIMALAYAN MOIST TEMPERATE FOREST: SUB-ALPINE SCRUB FORESTS, ALPINE SCURB AND PASTURES.

Thus, the flora of the Park can be broadly sub-divided into three forest zones:

(a) Temperate Broad-leaved Forest (1829-2730m).
(b) Mixed Coniferous Forests (2730-3650m)
(c) Alpine Scrub and Grasses (above 3650m)

The Temperate Broad-leaved Forests are dominated by *quercus Lineata, Q lamellosa, Q. pachyphylla, Schima wallich;;, Castanopsis spp,. Acer cambellii, Engelhardtia spicata, Juglans regia, Michelia cathcartii;, Magnolia cambelli, Cinnamomum obtusifolium* etc.

The mixed coniferous forests occur at higher altitudes with fir, *Abies densa , Maple, Acer spp.*, Spruce, Picea morindoides and junipers *Juniperous spp.*, with associates of taller varieties of Rhododendrons and Bamboo *Arundinaria spp.*

In the Alpine Scrub and grasses, the common shrubs and herbs found are. *Meconopsis spp, Primula spp.*, Dwarf Rhododendrons, *Rheum spp.*,Prostrating junipers, *Iris gentiana, Anemone, Delphinimum* etc.

Besides, there are many medicinal herbs found in the Khangchendzonga. *Aconitum spp., Picrorhiza kurrooa, Orchis latifolia, Nardostacl,ys jatamansi, Rheum emodi, Panax pseudogisenq* etc. are a few to mention here. Flora has been discussed in detail in previous chapter.

Fauna

Khangchendzonga National Park is a reservoir of gene pool with diverse habitats. The faunal wealth is also equally rich in content. The high altitude alpine and plateau regions harbour rare and endangered species of animals. The Snow Leopard of the alpine land holds the position at the apex of the biological pyramid. Himalayan Red Panda (State animal), Musk deer, Nayan or the Great Tibetan Sheep, Bharal or Blue Sheep, Himalayan Tahr- the Shapi of Sikkim, Marco Polo Sheep (reported), Leopard, Goral, Serow, Barking Deer, Lesser Cats, Wild dogs, Tibetan Wolf, Mountain Fox, Tibetan Fox, Himalayan Black Bear,

Marmots and Monkeys are a few principal animals found in the Park.

Avi-Fauna

As the State is rich in bird life with around 550 species and sub-species, the Khangchendzonga also harbours many forms of avian beauty and rarity. The flamboyant high altitude pheasants which include Monal Pheasant, Tragopal Pheasant and Blood Pheasant (State Bird), Tibetan Snow Cock, Himalayan Snow Cock, snow Partridge, Hill Partridge, Lammergier, Forest-Eagle Owl, Tibetan Horned Eagle Owl, Eagles, Falcons, Gawks ,Snow and Rock Pigeons adapted physiologically for those special climatic conditions at high mountains add to the number of smaller bird-lives found in the National Park. They plumet down to lower elevations in winter only when the upper reaches are covered with snow and most vegetation has died down. The high altitude lakes form stop-over sites for some of the migratory water fowls viz Bareheaded Goose, Eastern Goosander, Brownheaded Gull, Pintail, Grebes and Blacknecked Crane. These lakes are also the breeding grounds of some of the resident waterfowls viz, Brahminy Duck, avocet, Coot, Shovellers, Pochards are a few to mention here.

MOUNTAINS AND PEAKS IN THE KHANGCHENDZONGA (High Altitude) NATIONAL PARK, SIKKIM

	MOUNTAINS	*HEIGHT IN METRES*
1.	Mt Khangchendzonga	8585
2.	Mt Narsing	5825
3.	Mt Siniolchu	6886
4.	Mt Simvo	6811
5.	Mt Pandim	6691

	PEAKS	*HEIGHT IN METRES*
1.	Tent Peak	7365
2.	Talung Peak	7349
3.	Goechala Peak	6115
4.	Jhopuno Peak	5936
5.	Paki Lho Peak	4144
6.	Lamo Angdang Peak	5868
7.	Singdamringu Peak	3751
8.	Lhokhamburich	5497
9.	Yajuknamteng Peak	5643
10.	Kabur Noth Peak (or Kabru)	7338
11.	Kabur South Peak (-do-)	7317
12.	The Twins	7350
13.	Nepal Peak	7150

GLACIERS AND LAKES IN THE KHANGCHENDZONGA (High Altitude) NATIONAL PARK, SIKKIM

GLACIERS	*LAKES*
1. Zemu Glaciers - 26 Km long and about 300 m wide, largest in the area.	1. Lhonak Tso
2. Tent Peak Glacier	2. Green Lake
3. Nepal Gap Glacier	3. Langpo Lake
4. Zumthul Phuk Glacier	4. Dudh Pokhari (Ome chu)
5. Tongshion Glacier	5. Bhale Dudh Pokhari
6. Talang Glacier	6. Nor Pokhari
7. Siniolchu Glacier	7. Sungmoteng chho
8. Simvo Glacier	8. Tikuchia Pokhari
9. South Simvo Glacier	9. Majur Pokhari
10. Twins Glacier	10. Leduwa Pokhari
11. Hidden Glacier	11. Dalle Pokhari
12. Thekang Kiong Glacier	12. Lam Pokhari
13. Changsang Glacier	13. Tinkone Pokhari
14. East Langpo Glacier	14. Sukhe Pokhari
15. South Lhonak Glacier	15. Lamgepui Chho
16. Jonsang Glacier	16. Kishong Chho
17. Onglakthang Glacier	17. Shingo Chho
18. East Rathong Glacier	There are several other lakes

NATURE TRAILS

Trekking

Trekking in Upper Rangit Basin of Sikkim can be a wonderful experience-an antidote to the stressed city life. While trekking one relishes the pure air and silence of the wilderness as one walks through luxuriant forests and landscapes and pass through quaint villages. Although every nook and corner of Sikkim from the sultry tropical forests in the south to the wilderness in the north is a trekkers' paradise, some of the popular treks that are identified are (Rai et al[16], 1998):-

(a) *Monastic Trek.* Pemayangtse- Sangacholing-Khecheoplri-Dubdi -Sinon -Teshinding- Ralang. (March to May and October to December).

(b) *Rhododendron Trek.* Nayabazar-Hillay/-Barsay-Dentam-Pemayangtse. (March to May).

(c) *Khangchendzonga Trek.* Yuksom-Bakhim-Tsokha- Dzongri-Thangsing-Zemathang/- Goechala and back. (Mid March to mid June and September to (November).

(d) *Coronation Trek.* The first king Phuntshog Namgyal trekked from Gangtok-Rumtek-Sang-Yangang- Ravangla-Tashiding-Yuksom, and this trek has been established since then. The best time to trek through this route is October to December.

The Department of Tourism is planning to open trekking in the following proposed circuits in the study Area :-

(a) Ravangla- Maenam-Yangang-Simchuthang-Sirwani.
(b) Ravangla-Maenam-dadeillydaara-Borong/Sada.
(c) Rabdentse-Pemayangtse-Sangacholing-Khecheopalri-Norbugang-Dubdi-Tashiding.

MAN-MADE ASSETS
MONASTIC SPIRITUAL CENTRES

There are no remains of great archaeological interest in the study area except Rabdentse, but some of the Buddhist monasteries are worth visiting. The monasteries of Sikkim apart from being the places of religious discourse, worship and meditation are the storehouses of many hand written religious books and murals, the part and parcel of Sikkimese cultural heritage. Some important religious places are given in succeeding paragraphs.

Pemayangtse Monastery. The sublime Lotus monastery lies North of Gayzing. Built in 1705 it is Sikkim's second oldest monastery. This ancient monastery is the headquarters of the Nyingmapa Buddhist sect of Sikkim. Inside are great murals filled with thousands of deities from the Tibetan Buddhist pantheon. On the top floor is a wooden sculpture, the Zandog-palri, which depicts the heavenly abode of Guru Padmasambhava. The main ritual masked dances of this monastery are held annually during the month of January.

Dubdi Monastery. Built by Gyalwa Lhatsun Chenpo in 1642, this is the oldest Monastery of Sikkim situated on a hill top above Yuksom amidst a verdant tranquil forest. An hours walk uphill from Yuksom passing through cardamom fields, and lush Green forest, the walk is well rewarding. The monastery has been able to retain its old design and architecture with beautiful wall paintings and old thankas.

Tashiding Monastery. Located in its majestic isolation on a helmet shaped conical hilltop, this monastery belongs to the Nyingmapa sect. Built in 1716, according to legend, Guru Padmasambhava shot an arrow and meditated on the spot where it had fallen. Tashiding was constructed on the spot, surrounded by deep forest and two rivers, the Rathong and Rangit. This monastery houses the sacred chorten.

Thongwa Rangdol Which according to belief, just a glimpse of the holy chortens cleanses one of all sins. The sacred Bumchu festival is performed here on the 14th & 15th of the first Tibetan calendar. More details on Tashiding monastery are given in the chapter on Spiritual Ecology of Sikkim.

Ralang Monastery. The oldest monastery of Karma Kagyu sect in Sikkim is located at Ralang in the South Sikkim district. The full religious name of the monastery is Karma Rabdenling Gompa. The monastery was constructed in 1718 by the fourth Chogyal of Sikkim during 1717-1733 on his return from pilgrimage in Tibet. A new monastery has recently been reconstructed on a larger scale with a huge golden painted Statue of lord Buddha inside it. Old Ralang monastery is still functional but is in a state of slow degeneration.

Khecheopalri Lake. This lake is set amidst the thick and luxuriant forest cover in West Sikkim. Khecheopalri lake is considered as one of the most sacred lakes of Sikkim, and it is believed that wishes are fulfilled when prayers are offered to the deities Of the lake. Many domestic as well as foreign tourists visit this lake each year. The lake Attracts many migratory ducks during the winter season. The Khecheopalri monastery is Situated above the lake and is half and hours walk uphill. A large number of visitors come to this lake during the khecheopalri Lake Festival which normally falls in the month of March. Khecheopalri lake has been dealt with in greater details in previous chapters.

SECTION III
IMPACTS OF MOUNTAIN TOURISM IN UPPER RANGIT BASIN

Mountain Tourism

The mountain people's daily needs are met through resources found in the Himalayas. Local demand for these resources is growing at a fast rate and thus leading to gradual over harvesting and environmental deterioration. Local people have no option but to continue to use the resources because development has not been able to mitigate poverty and generate new employment opportunities in remote mountain areas. Poverty mitigation in the mountain region requires accelerated use of the resources and this has led to rapid environmental deterioration. The search for sustainable livelihood options under these conditions lies in harnessing natural comparative advantages. Tourism in the Himalayas has been gaining in popularity, and it has provided new opportunities and challenges. It has been able to provide income and employment opportunities to remote and relatively inaccessible areas. Mountain tourism has both positive and negative impacts.

Tourism has affected traditional land-use practices and brought about changes in crop cultivation and cropping patterns. Conversion of land forest to agricultural use, conversion of agricultural land to build lodges or tea stalls, leaving land fallow to rent out as camp grounds, etc are some of the changes brought about by mountain tourism. In some places, agriculture has become a second occupation to tourism-related activities. The main impacts of mountain tourism changes in land use, garbage collection, littering, pollution, and increased demand for firewood.

Mountain tourism creates jobs to mountain people annually in the form of porters and other staff. Also, many women are employed in lodges and tea houses. The jobs created by tourism in mountain areas, however, are seasonal.

Wages paid to the porter accrue as income. But all income may not be retained within mountain areas as porters may come from other regions where tourism is not active. In the lodges, income accrues from accommodation and food payments. Income leakage occurs as many items have to be imported to supply food and fulfil the needs of visitors.

Main Impacts

The impacts of mountain tourism may be grouped into three main headings. The impacts may be either positive or negative. It is not always possible to discern whether the impact is positive or negative, since its true manifestation may take many years.

Also, tourism may not alone be responsible for the impact as many other factors may be including changes in the mountain environment, economy, and sociocultural practices of the mountain people. The main impacts witnessed in the case of the mountain areas of 'Upper Rangit Basin' are briefly highlighted :

(a) Environmental.

(b) Economic.

(c) Sociocultural.

Environmental Impacts

Land Use Patterns

(a) **Changes in Cropping Pattern**. The areas adjacent to Pemayangtse and Pelling, the areas on the road joining Ravangla-

Sosing-Kewzing-Legship and the areas on the Tashiding salient upto Yuksom have maximum physical and social contacts with the tourists. The locals in these areas have learnt that cash crops bring quick cash returns to the house holds, thus motivating change. Potato, fruit and vegetable cultivation in the Home Gardens has increased replacing traditional staple crops such as 'buckwheat' or barley.

(b) Changes in the forest encroachment are mixed. The dense Maenam forest gets thinner by the year, giving space to new buildings. Within a period of three years from 1997 to 2000, a 1.5 km stretch of forest area along Ravangla- Ralang road had given way to human habitations resulting in forest clearance upto a depth of about 200-400 meters along the stretch. Similar thinning of forest was noticeable on the Western Pelling spur, and in the area around Sinon Gompa located North of Tashiding Market.

(c) The land on the Yuksom-Dzongri corridor finds a different kind of change in land use pattern. Land is left fallow to provide camping grounds replacing cultivation of traditional staple crops.

(d) In Yuksom-Dzongri-Tsoka-Bakhim areas, pack animals are gaining ground compared to mulch animals, as they bring higher cash returns to households by carrying tourists loads. This may later result in added demands for fodder and grazing areas at the cost of forest land and wildlife.

Litter, Garbage and Pollution

Another major environmental problem created by mountain tourism on the Rhododendron trails and the Yuksom-Dzongri corridor region is the accumulation of the solid waste including human waste along the trails, in the campsites and in and around villages. Independent trekkers and local people are the main sources of littering. The organised trekking groups litter the campsites if their trek leader is not an environmentally conscious person. Porters and other local support staff also litter trails. Lack of awareness among the local villagers means they just throw all their rubbish in backyards or in the river or in the street, where it is spread around by wind or scavenging dogs, cats, rodents and birds.

The main components of solid waste found along the trail, at campsites and in the villages are mineral water bottles, batteries, toilet rolls, toilet paper, wrappers, metal cans and glass bottles.' Visitors buy bottled mineral water from Yuksom, Gayzing, Pelling or Ravangla for their own safety but leave them behind on the trails or in villages. . If the bottle is not damaged, it can be reused, but once it is damaged, it ends up as rubbish.

Estimates indicate that an average trekking group of 15 people generate about 15 kg of non-biodegradable, non burnable waste in ten trekking days. With thousands of trekkers in the mountains, the waste multiplies (Banskota and Sharma[17], 1998).

Litter on the wayside is ugly and unhygienic. Time taken for various items to decay as given by ICIMOD[18] (1997) is as under:

(a) Traffic tickets	- 2 - 4 weeks.
(b) Cotton rags	- 1 - 5 months.
(c) Ropes	- 3 - 14 months.
(d) Wool sock	- One year.
(e) Bamboo pole	- 1-3 years
(t) Painted wooden skate	- 13 years
(g) Tin can	- 100 years.
(h) Aluminium can	- 200-500 years
(j) Plastic covers	- 450 years.
(k) Glass bottles	- Undetermined

Forests

Three factors have increased the demand of firewood:-

(a) Tourists outnumber local people on the Rhododendron trails and Yuksom-Dzongri corridor.

(b) Although group tourists and independent trekkers are expected to use alternative sources of energy, they and the porters depend on firewood.

(c) Growing seasons in higher mountains are extremely short.

Economic Impacts

Economic impacts of mountain tourism can be viewed in terms of how different aspects of the mountain economy are affected by tourism expenditure and tourism development. The full impact of mountain tourism on mountain areas will depend on how strongly the tourism sector is linked to the mountain economy. (Banskota and Sharma[19], 1998).

The current flow of tourists in the Upper Rangit Valley is not too large, and has no immediate implications on the local production system.

Arts and Crafts

The one sector that has particularly benefited is the Sikkimese arts and crafts :

(a) Handicraft Centre, Tibetan Settlement, Ravangla

(i) The carpet weaving centre with looms of all sizes has grown in size and local sales have picked up. The carpets are of high quality, especially, those woven by Norwegian wool.

(ii) Wood carving centre products sales have been popularised.

(iii) Sale of Tibetan arts products, including literature have also picked up.

(iv) Kewzing handicraft centre sales have picked up.

(b) Hotels and Resorts

(i) Between 1997 and 2000, eight new tourist lodges have come up in Ravangla, of these, one tourist lodge 'Narsing Village Resort' is doing a booming business. It provides ethnic village kinds of facilities, though in Western style, and draws the richer variety of visitors.

(ii) Barring *momo* and noodles, ethnic foods such as *Kinema* and *gundruk* are still not popular with visitors from plains.

(iii) Guided tours, with reservations done at Calcutta, Delhi and Siliguri, for lodging as well as transport are on the increase.

(iv) Fresh employment has been generated in the lodges for the erstwhile and now abandoned tea garden workers at Ravangla.

(v) *Thangka* painters have been provided opportunity for jobs and for exhibiting their products in Narsing Village Resort.

(c) Transportation

(i) The Sikkim Nationalised Transport (SNT) buses are not very popular with the tourists.

(ii) Luxury coaches, mostly continue to be hired form Siliguri for package tours.

(iii) The 'Town committee' of Gyalshing, in order to promote maximum use of local jeeps and taxis by visitors, do not permit outside taxis to ply beyond Pelling. As such visitors are forced to hire local jeeps and taxis for visiting Khecheopalri Lake and beyond. For this action the Gyalshing Town Committee have procured a positive nod from the Ministry of Transport, Sikkim

(d) Guides and Porters. The support staff in terms of guides and porters are required by trekkers and mountaineers on Rhododendron trails and Dzongri treks.The Sherpas and Limbus of West Sikkim are most benefited by this requirement. Trekkers, mountaineers and certain other tourists could have a greater impact on income in the Upper Rangit Basin. At present, in addition to what they import, group trekkers and mountaineers purchase most of their food items in Darjeeling and Siliguri prior to treks or expeditions. Many of these items, which include vegetables,eggs, milk and other perishable food, could be produced locally. Encouraging the production of these items at Legship, Gayzing, Ravangla, Kewzing, Tashiding and Yuksom would benefit these areas even more than indicated above.

Tourism does not generate spontaneous benefits. Tourism is unlikely to bring community development unless efforts are made. Tourism development in mountain areas must be integrated with community development. Indigenous knowledge alone cannot help promote tourism in mountain areas. New knowledge and technology will have to play a greater role. The economy of the Upper Rangit Basin cannot depend solely on food production. Alternatives need to be developed. The plans for the twenty first century must be based on comparative advantage paradigm. (Banskota and Sharma[20], 1998).

Socio-cultural Impacts

Mountain tourism in Upper Rangit Basin is likely to bring about and affect certain sociocultural changes in the local inhabitants of the area, such as :

(a) Change in people's behaviour, dress, lifestyle, family and social structure due to modernizing trends.
(b) Values and expectations.
(c) Decline in local support for traditions and institutions.
(d) Peoples preference for tourist-related jobs over education.
(e) Pollution of sacred places.
(f) Changes in traditional architecture.

Not all effects are likely to be negative. Although not directly attributable to tourism, there are some positive impacts that can be identified:s

(a) Poverty alleviation.
(b) Awareness generation. (Education, health and hygiene, conservation of natural and cultural sites etc.).
(c) Development of infrastructure, settlements, cottage industry etc.

(d) Research on nature and spiritual/cultural related aspects.

(e) Publicity for the region and Sikkim.

Research. Many research studies have resulted from tourism. The monasteries of Pemayangtse and Tashiding have drawn research scholars and scientists from India and abroad for deeper studies into Mahayana Buddhism and monastic systems.

International Publicity

The area has received a great deal of publicity through casual visitors and tourist. A number of books Sikkim, with details of Rangit Basin have been printed. Many articles on subjects relating to Khecheopalri Lake, Rathong chu imbroglio, Bum chu festival etc have been published in international journals. All of these have generated an awareness about the unique natural and cultural heritage, as well as the problems to be addressed in the Rangit Basin

Summary of Impacts

The main aspects, and negative and positive implications emerging from tourism in Upper Rangit Basin are summarized in Table No. 6.2.

References

1. Mridula and Narayan Datt, *Ecology and Tourism*; Universal Publishers Distributors, New Delhi, 1991.
2. Whelan, T., (ed), *Nature Tourism* ; *Managing for the Environment*, Island press Washington D.C., USA 1991.
3. Brandon, K., *Ecotourism and Conservation : A review of Key Issues*, Environment Department Papers, No 33, The World Bank, Washington D.C., USA, 1996.
4. Whelan, T., 1991, *ibid.*
5. Adventure Travel society, Press Release - Adventure Travel Congress, Colorado, 1998.
6. Brandon, K., 1996, *ibid.*
7. Lindberg, K, and hawkind, D.E., (eds), Ecotourism : *A Guide for Planners and Managers*, North Bannington VT., The Ecotourism Society, 1993.
8. Sharma, E., Jain, N., Rai, S.C., and Lepcha, R., Ecotourism in Sikkim : Contributions towards Conservation of Biodiversity Resources. In, D. Marothia, (ed), *Institutionalizing Common Pool Resources*, Concept Publishing Company, New Delhi, 2000.
9. ______________________. *ibid.*
10. Draft Ninth Plan. Government of Sikkim Document, Bureau of Statistics, Gangtok, 1997.

Positive/Negative Impacts of Tourism-Upper Rangit Basin

Negative Impacts			*Positive Impacts*		
Ecological/ Environmental	*Sociocultural*	*Economic*	*Ecological/ Environmental*	*Sociocultural*	*Economic*
Destruction of :- .Forests (energy) .Fauna(poaching) .(Grazing land (camping)	.Cultural loss .Alienation .Xenophobia .Imitation of western culture	Encourages :- .Inflationary trends conspicuous consumption .Demonstration effect. Imports for Tourism/leakages treatment of tourist.	Encourages .Ecological awareness. .Conservation measures. ..Measures to hold pollution. .Cleanliness in camp grounds. Park/woodlands.	Preserves :- .Cultural monuments. .Folk traditions. .Cultural properties .Art and History	Creates :- .Job opportunities. .Foreign exchange. .Additional income/tax. .Better multiplier effect .Diversification of economy. .Job for unskilled/ semi-skilled
Pollution:- .Water .Air .Noise .Trash, garbage, trails, etc	Changes:- .Lifestyles .Native architecture. .Settlement patterns. .Folk traditions .Customs.	Withdraws:- .Labour .Overburdened communal services. .Overuse of sacred resources. .Uneven economic development. .Tourist enclave.	Helps :- .Maintenance of scenic Landscape. .Research/ environmental impact studies. .Retreat from marginal hillside farming.	Responsible :- .Tourist .Space organization. .Urbanisation. .Increased communica-tion. .Modernisation of facilities.	Improves :- .Infrastructure. .Local arts/crafts. .Regional development. .Standards of living.

Degradation of :- .Landscape. .Scenic appeal. .Promotion of throwaway mentality. .Congestion/ overburdening. .Hygiene problems.	Promotes :- .Moral laxity. .Crime/ gambling. .Prostitution. .Drug addiction .Inferiority (locals) .Beggar mentality. Disturbs :- .Religious practices. .Indigenous style .Social cohesion .Promotes neo-colonialism.	.Seasonal dependence. .Dependent on tourism. .External domination (city).		Encourages :- .Education. .Training of craftsmen/ expedition. .Contact with outside world .Speedy knowledge. .Rediscovery of lost traditions. .Fund mobilization.	

Source: Singh, T. V. The Kullu Valley: Impact of Tourism Development in the Mountain Areas. Study supported by ICIMOD; Himalaya Books, 1989

11. ____________________.*ibid.*
12. Rai, S.C., E. Sharma, R. Lepcha and R.C. Sundriyal, Tourism Development in Sikkim, In *Sikkim : Perspectives for Planning & Development*, Sikkim Science Society, Tadong, 1998.
13. __________________. ibid.
14. Maharana, lyatta., GBIPHED, Tadong, 2000.
15. Forest Department, Government of Sikkim., *Khangchendzonga* (*High Altitude*) National Park, Wildlife Circle, Big Byte Computer. Systems, Lotus Printers, Gangtok, 1998.
16. Rai, S.C., E. Sharma, R. Lepcha and RC Sundriyal, 1998,)ibid.
17. Banskota, K. and B. Sharma, *Mountain Tourism for Local Development*, ICIMOD and CREST, Kathmandu, Nepal, 1998.
18. ICIMOD, *Manual of Rural Technology with Implications for Mountain Tourism*, ICIMOD, Kathmandu, Nepal, 1997.
19. Banskota, K. and B. Sharma, 1998, *ibid.*
20. ——————————————*ibid.*
21. Singh, T.V., The Kulu Valley: *Impact of Tourism Development in the Mountain Areas*, ICIMOD, Himalayan Books, 1989.

7

Understanding Human Ecology of Upper Rangit Basin : Some Reflections on The Peoples Psyche

SECTION-I
INTRODUCTION

General

This chapter is intended to present a holistic view, based on the study and research carried out within the precincts of human ecology in the previous chapters.

The social structure of the Rangit Basin inhabitants is based on social status ascribed mainly by tribal and non-tribal affiliations. The Bhutia and Lepcha communities, though economically better off, enjoy the status of being scheduled tribes, giving them additional constitutional liberties. The internal developments and exposure to the world have tended to bring about some social and political changes. These changes can be understood in the context of traditions and modernity. In fact, it is seen that even most other states in India live in conventional conservatisms, despite the fact that modernity dominates them. The people of the Basin have a keen desire to grow modern. This desire, especially amongst the young is at the cost of losing out on their traditional knowledge of ecology and customs gathered over generations of learning, centuries of experience and interrelationship. As regard the Upper Rangit Basin, the interplay of modernity and tradition reveals the following propositions:-

(a) Intra- contradictions of traditions of the society.

(b) Inter- contradictions of the society.

(c) Intra-contradictions within the modernity in the context of the society.

A mutual incompatibility and antagonisms between tradition and modernity are latent, and the balance is judged and cleared between modernizing invasive forces on traditional forces and vice-versa. The variation between changing values is seen in cultural, socio-political, ethno-biotanical, education and occupational scenes.

This affects the ways of thinking of the people and generates conflicts between self, society and the habitat.

The Isolatory Setting.

The people residing in the Upper Rangit Basin have lived a generally secluded life for centuries. The folk groups of the area have therefore enjoyed a very close relationship with their environment. This closeness is explained by the fact that these people have lived on this land, gaining their livelihood directly through such primary activities as farming, herding, hunting, gathering and fishing. A great many facets of Upper Rangit culture relate at least indirectly to the local ecology and geographical setting, and involve adaptive strategies. The languages of the three major folk groups bear the vocabularies required to exploit the habitat. Their religious acts are required to mitigate their environmental hazards, their folk tales honour great hunters shamans and lamas, their proverbs offer wisdom concerning the weather and proper time for planting, and their traditional architecture reflects the environmental determinism caused by the available local building material and the climatic condition.

While the Kanchandzonga Massif in North and less relatively the Maïnam-Tendong ridge to the East affords them geographical isolation; the political isolation comes from Singalila range forming the political boundary with Nepal.

The Historical Background

History of Sikkim is conspicuous by the fact that, one does not find a single mention of the Sikkimese armed groups having ventured out to trans border areas to capture adjacent lands. They have felt self sufficient within their setting. They have however been attacked upon on several occasions as cited below:-

(a) The Tibetans of U Tsang and Kham province found their way into Sikkim through the Northern and Eastern passes. They came in search of agricultural and pasture lands, till they finally established the Kingdom under the Chogyal at Yuksom in year 1641, displacing the Lepchas.

(b) The Nepal Gurkhas conquered the entire Rangit Basin and annexed areas upto Tista including Darjeeling by the year 1814. Meanwhile Bhutanese excursions from East resulted in loss of Kalimpong, and other areas West of Dongakiala- Chola range.

(c) The British intervention helped regaining the lost areas at the cost of permanent loss of Darjeeling and Kalimpong to the British Raj.

(d) Democratisation lead to the political merger of entire Sikkim kingdom with the mainland India.

The history clearly indicates that the Sikkimese continued to live secluded and relative militarily passive lives within themselves with continued threats from their neighbours in North, East, South and West.

Features Affecting Ways of Thinking

There is a lot in common between the mainland India, and the state of Sikkim, that have guided their cultural phenomena and ways of thinking:-

(a) Both have faced aggressive neighbours, who through armed invasions captured their entire lot and established their empires.

(b) Both assimilated their invaders into their own folds, till the invaders themselves became an integral part of the Indian and Sikkimese cultures.

(c) The cultural diffusion occurred at both the places, giving way to modified cultures with the progress of time

(d) Religions have been a major force as guiding principles for their ways of thinking.

The major difference between the two were however their size, and their geographical settings. While in mainland India, there was a multiplicity of cultural entities, Sikkim due to its size and isolatory setting was able build a culture of its own.

Certain other features affecting the cultural buildup, and thought processes within Sikkim are:-

(a) The continuous unhindered rule of the Chogyals, whose writ as a monarchy system was absolute.

(b) The pronouncement of Mahayana Buddhism as the state religion, and thereby making the monasteries and Lamas nodal points of power and economy.

(c) A thin density of population, a total lack of infrastructure and almost non- existence of large industries lead to cultural diffusion of population within the isolated pockets.

(d) The absolute dependence on their lands for subsistence, resulted into a store house of thorough and detailed indigenous ecological knowledge of fauna and flora, and other environmental elements.

(e) The local indigenous arts and crafts developed, which, although they had strains of Tibetan, Nepal and Indian cultures, did have a stamp of originality.

(f) Education remained confined to monasteries. Schools beyond primary classes were few and far in-between. Secondary and post secondary education was confined to a few larger townships such as Namchi and Gangtok.

(g) Folk medicines, including shamanic medicinal practices became a way of life, and the belief in lamas, shamans and '*Jhakris*' became almost absolute.

The Issues

The information technology, and inter actions with the society of popular culture, so also the conflicting legislations in terms of job reservations have had telling effects on the ways of functional of the people. The succeeding section discusses these issues.

SECTION II
ASSIMILATION OF WAYS OF THINKING

Ways of Thinking : Conceptual Aspects

Our sense of belonging to one world has never been keener then at present. Yet the emphasis today on this evident fact itself implies that while every individual is affected by the quickening flow of the world events, he is still strongly influenced by the ways of living and thinking in his own society and culture, and his religious realms.

The phrase 'ways of thinking' refers to any individual's thinking in which the characteristic features of the thinking habits of the culture to which he belongs are revealed. 'Ways of thinking', as used in this chapter designate value judgments, values in ethics, religion, aesthetics and other concerns of human ecology. Therefore, the 'ways of thinking' of a community are conditioned by the cultural habits and attitudes while giving vent to their thoughts and applied actions.

The conclusions on the ways of thinking have been drawn by the method of holistic study of human interactions in previous chapters, furnishing the bases for this study.

As studied in previous chapters, there is a mention of fusion of cultural and religious traits between the Hindu Nepalis, the Buddist Bhutias and Lepchas and the Animist tribals- the Kiratas. An insight needs to be furnished by the way one group of people adopts the ways or patterns of thinking of another group. One community does not generally adopt the ways or patterns of thinking of another culture straight away. On the contrary it begins with:-

(a) Criticism of the 'alien' ways.
(b) Selective adoption of the borrowed cultural traits.
(c) Modification of traits during the course of adoption.

Modification of Buddhism in Sikkim

Buddhism in its original form was the one preached by Gautama Buddha after enlightenment, and his subsequent preaching to the 'Sangha', commencing with the first preaching of the 'Four Noble Truths' at Sarnath. From the point of view of Sikkim, modification took place in following order:-

(a) Amalgamation of Upanishads by Nagarjuna to the initial texts of original Buddhism as preached by Buddha himself.
(b) Formulation of Vajrayana through the fusion of Tantricism, thereby by initiation of the Karma-Kagyu sect, which is one of the two dominating sects in Sikkim.
(c) Spread of Mahayana Buddhism in Tibet, and its modification through fusion with the original Bon faith existing in pre-Buddhist Tibetan society.
(d) Reformed sects within modified lamaist Mahayana Tibetan Buddhism.
(e) The rise of the Lamas, and centralization of political and religious power by the reformed school of Gelugpa Sect in Tibet.
(f) Tibetan immigration to Sikkim by majority of unreformed school of the original Nyingma Sect and subsequent Karma Kagyud sect.
(g) Fusion of Tibetan Buddhism with the animist Mon and Bonthin constituting the Lepchas, and to a lesser extent the Kiratas (Limbu, Rai and Gurung).
(h) Fusion with strains of Brahmanic Hinduism from India and Nepal into the monastic education systems in Sikkim.

The adaptations within religious realms did not occur through diffusion of people alone. Nature and the natural objects played their own part. Historical events and terrain configuration conferred piety and sanctity to natural terrain features and to locations where monasteries were constructed. As such, in accordance with the wisdom of the Lamas, Kanchendzonga became the diety goddess of the population. Rathong chu became the sacred river, and the water of Rathong chu began to sanctify the entire population of Sikkim during the festival of Bum chu at Tashiding. Myths and folklores were added, and attached to religious rites. Khecheopalri became the wish fulfilling lake. All other mountain peaks, rivers and streams were given their respective cultural and religious sanctity. Nature, history and religion fused together to become a part of culture of Upper Rangit Basin, giving its inhabitants their own 'ways of thinking' and formed the basics of current status of human ecology in the basin.

Certain Reflections on Sikkmese of Upper Rangit Basin 'Ways of Thinking'

1. **Literary poems on the Dead**. The people in the Basin live in scattered villages and pastures connected by narrow routes. To deviate from these paths is to risk death in a mountain wilderness. Having to live under such severe conditions, the people are confronted with the physical threat to their existence and are constantly faced with the problems of survival. Death is a favourite subject in their poetry:-

(a) The Hindus derive the poems relating through the God of Death-*Yamraj*.

(b) The Bhutias sing the gloom and despair through the poems of Tibetan Monk Milarepa, the religious poet.

2. **Plural Marriages.** Polyandry, especially amongst the Bhutias and Lepchas was a way of life till recently. This system has however slowly faded out. The principle of polyandry was based upon adaptation of latent mobility to pasture lands (nomadism) on one hand, and an attempt to heighten the stability of a household on the other.

3. **Remarriage.** These days most Lepchas and Bhutias adhere to monogamy. They however remarry immediately after the death of their partners, and as a result, there are virtually no widows or widowers on the society, barring the Chettri and cetain Brahman castes of Nepali Hindus.

4. **Festival Celebrations**. Sikkimese dancing is exciting and inciting. On the major festival days they dance and consume '*chhang*' and

'*raksi*' round the clock. This is more pronounced during the '*Losar*' festivities (Sikkimese new year) when dancing drinking and feasting continues for the entire week. A public holiday is declared during the period, this period and all government institutions remain closed.

5. **Becoming a Lama.** Amongst Bhutia and Lepcha families, generally at least one among five children goes under the lamaist order. They join one of the larger monasteries for religious and monastic education. This subsequently results into a non -functioning (other than religious rites) population to the extent of 20 percent of the total whole within these two communities.

6. **Power of the Lamas**. Due to total faith by the followers, the people accept calmly what would seem irrational to the eyes of modern people. Lamas are said to have powers to stop or bring rains, cease or stop earthquakes and also to stop invading armies of an alien enemy.

7. **Prayers and Preaching**. The congregation in the monasteries on prayer days are large. The lama leads the choir singing of prayers, and entire congregation follows suit. The meanings of the prayer hymns are however not preached. These sacred secrets stay with the lamas alone. The locals therefore do pray, but what their prayer means they do not know. Hindu temples also display a similar phenomena where all '*Shlokas*' are in Sanskrit. Even during marriage ceremony, the rituals are in Sanskrit, and the couples simply feel happy that the rituals have been completed. The local ways of thinking are adjusted towards formalizing of a ceremony rather than understanding of the underlying message in prayers.

Way of Life and Personal Attributes

Lack of Competition. There are seat reservations at national level for Sikkimese for higher studies in medicine, engineering and other subjects. By virtue of reservations, students scoring low pass marks also get admissions in prestigious institutions. This ultimately reflects on their professional attributes in the negative. Field study showed that the local populace preferred non Sikkimese doctors to their own, as they had reservations about their professional abilities. This relaxed attitude needs to be converted to positive action and active 'way of thinking' for future generation.

Personal Discipline. A glance through the police records at Ravangla and Gyalshing reveals that barring some stray incidents of burglary; the state of discipline is clean. This clean mannerism is visible on the roads and in the market places. Fights are uncommon. Although there is a great deal of consumption of alcohol, the drinks related fights are generally not heard of. Here the high satisfaction level of the people is revealed. It also indicates the mutual love and respect within the societal groups.

Sports and Pas times. Baichung Bhutia, the finest Indian footballer belongs to Ravangla in the Upper Rangit Basin. Young boys and girls actively participate in games. The most popular game is football, followed by volleyball and basketball. The people here have tough strong and flexible body structures, and if trained well, they may turnout to be excellent gymnasts, and divers. Many locals, especially sherpas specialize in mountaineering. Carom-board is a popular indoor game. Betting in games is common. It is generally seen that a Sikkimese, when in an up beet mood, is capable of betting on anything. This indicates happy frame of mind. More importantly it indicates the capacity of the people to lose or win at any one instance, and to recommence buildings up of their fortunes from 'zero point'.

Nonviolence. Though most people consume meat, they abhor violence against men as will as other animals. During period of research, while driving down the roads, one encounters large numbers of snakes crawling on the roads. The local drivers ensure that no snakes come under their wheels. Also, unlike elsewhere in some parts of India, the street dogs are seldom or never stoned. As a habit therefore, the street dogs themselves are not aggressive . This reveals the harmony within the human society, and the sense of responsibility towards all other living beings on the biosphere. This is a practical practice of 'Deep Ecology' and bio-centricism.

Conclusion

Above is an outline or a brief commentary concerning the ways of thinking, and action oriented attributes of the residents of Upper Rangit Basin of Sikkim extracted through personal contacts and other resources. These are a fall out from their lifelong experiences amongst themselves and with the nature gathered through cultural and religious attributes.

Selective adaptation to modern life styles will need to be done, mainly to ensure that the present and future generations of the Sikkimese do not find themselves left out of the mainstream culture and competition in the society as a whole. While retaining the virtues of their 'ways of thinking' , the Sikkimese need to guard against the vices.

8

Conclusion

SECTION - 1
SUMMARY AND FINDINGS

The study constitutes of eight chapters in all relating to a holistic assessment of the human ecology of the study area. Care has been taken to maintain a reasonable sequence in organizing the chapters. The study begins with the introductory chapter that encompasses a brief description of the problem in the context of human intervention into the natural environment. The chapter also includes aim and objectives of the study, hypothesis, methodology, chapterisation and brief review of the related literature.

Chapter 2 deals with the geographical setting of the region. The first part covers the details of physical geography of the areas under investigation i.e. the Upper Rangit Basin, and is followed by the second part that highlights the human geographical mosaic of the entire Sikkim state. The main findings could be summarized as follows :-

1. Entire Sikkim state forms the catchment of river Tista. River Rangit is major tributary that originates from Kanchendzonga massif and flows through West and South Sikkim, uniting with Tista at Melli confluence point.

2. The confluence of Rathong chu and Rangit chu, 3 kms upstream from Legship forms the Rangit River. The catchment area North of Legship comes within the confines of the Upper Rangit Basin, and to the South, the lower basin. The Kanchendzonga Massif to the North, and the Singalila and Mainam-Tendong ranges to the West and East further define the study area. The Upper Rangit Basin covers an area of 979.02 sq. kms.

3. Elevation ranges from 600 m at Legship to 7338 m at the highest point within the catchment. Average rainfall is over 2620 mm annually.

4. 38.96 per cent area within the catchment has a dense forest cover. In the lower regions agricultural lands predominate and the dense forests

are found in fragmented pockets. Agriculture activity is limited to altitudes below 3000 m, covering 160.88 sq km contributing to 16.44 per cent of total catchment. The upper regions of Alpine pastures and scrub land contributes to 117.72 sq. kms, beyond which is the zone of permanent snow.

5. Geologically, the basin comes under pre cambrian rocks belonging to Khanchendzonga group within Chungthang formation group. Daling phyllites are prone to landslides.

6. Sikkim has three main ethnic groups viz the Lepchas who are known to be original inhabitants, the Bhutias originating from Tibet and Bhutan, and the Nepalis. The Kirata groups dominate the population in the Basin living side by side with Bhutias, Lepchas and the other Nepali groups. The *lingua-franca* of the inhabitants is Nepali, though the various ethnic groups in the region have their own dialects.

7. Although a changing life style particularly in dress is observed among the people never the less the traditional dresses predominate in the rural pockets. Different ethnic groups have different dressing patterns for men and women. Bhutia married women wear a striped apron *(pangden)* to define their martial status. Nepali men may be recognized by their typical caps (*birkhe topi*).

8. Wearing of jewellery is common and is traditional. The Bhutia women follow the Tibetan pattern, while the Nepali women follow the traditional Hindu pattern. Ear rings, bracelets, necklaces, nose rings and brooches in respective ethnic styles are worn by women.

9. Rice is the staple food, and 'chhang' and 'raksi' the traditional alcoholic beverages. Bhutias prefer the salted butter tea, and other fermented and non-fermented food items of Tibetan origin. Bhutias and Lepchas prefer beef. However, the Nepalis do not take beef on grounds of religion. '*Dal-bhat-tarkari-achar*' is the staple Nepali diet.

10. Bhutia's and Lepcha's houses follow a two-storeyed rectangular pattern, where ground floor is mainly used for domestic animals. Split bamboo, thatching and stone masonary are all used. Nepalis have separate shelters for cattle. Concrete buildings are gradually coming up and seen to be preferred by almost all ethnic groups in the regions as modernity is fast penetrating into the rural country side.

11. Festivals are religion based, though all ethnic groups participate in festivities of each other. Tamangs celebrate both, Hindu and Buddhist festivals. Losar, Bumchu, Sagadawa, Tihar and Dasain are celebrated with religious fervour. Mt. Khanchendzonga, the guardian diey is worshipper during Pang-Lhabsol.

12. Development activities are a part of the five year plans. Agriculture is the mainstay of economy. Rice, ginger and cardamom are the main crops grown in the region followed by maize. Horticulture is quite popular. Oranges, banana and apples are grown in large scale. Most industries are either of small scale or medium scale.

The mountaineous terrain and high precipitation typify the physical features and climate. The existence of large number of water channels facilitate construction of mini to medium sized hydel projects. This has resulted in erecting of the Rangit Hydel Project. Electricity is available in almost all houses. The dense forests, rich in biodiversity lend themselves to traditional ethnobotanical practices, especially pertaining to arts and crafts, and to medicinal plants.

Soil in the region is rich in nutrients and the availability of rich bio-mass reduces dependence on chemical fertilizers which is a healthy sign. The loose soil on the slopes leads to frequent lands slides, and avalanches in the higher regions. Watershed soil treatment plans therefore need to, be given a high priority.

The literacy rates, especially of the girl students need a deliberate insight. PHCs, though they have requisite medical stores, hygiene and sanitation is below par. Availability and presence of trained medical doctors and nurses have to be ensured.

Quality of life overall is reasonably satisfactory, and will improve through education, and maintenance of traditional cultural and ecological knowledge.

Chapter 3 encompasses study of various aspects of religion which is an integral part of culture. It involves a set of beliefs and practices, which has given rise to a social system through which the people seek spiritual and physical harmony with nature, life and death. A summary of spiritual life, and its essential attributes to local culture within the realms of human ecology is elucidated below :-

1. Hinduism and Buddhism, both interlaced with animist spiritualism are the main religious traditions in Upper Rangit Basin.

2. Sikkim was under the rule of Chogyals until 1975, when Buddhism was the declared state religion. Though followers of Hindu faith are numerically larger, the state character is mainly Buddhist, which is evident from the number of monasteries, the architectural designs and the general way of life. The reservation of one seat for 'Sangha' in the state council further emphasizes the point.

3. The lama's word is taken as law by the inhabitants. The involvement of the lamas is found in all actions and activities including the rites during

births, marriage, death, agricultural sowing and harvesting, and celebrations of festivals.

4. The Upper Rangit Basin is although a small geographical area of less than 1000 sq kms, it contains the oldest and most important monasteries from all sects (except Sakyapa Sect) including a Bon monastery. Hindu temples, though smaller in size, and located at less prominent places, exist in all villages and towns.

5. The sites of monasteries is selected with great care, with due considerations to auguries and omens. The ground configuration and soil study are carried out prior to erection of structures. Similar considerations are followed for erecting *chortens*.

6. All monastries' entrances face East wards, while the altars are at the Western ends.

7. The government buildings and houses of prominent citizens have adopted many features of the monastic architectural designs, especially with respect to roofs, doors and windows, and the beams and pillars.

8. The prayer wheels and prayer flags are an integral part of monasteries, important buildings and all other habitats. Holy messages inscribed on the flags are believed to bring good luck and harmony.

9. The *chortens* contain and preserve relics and bodily remains of *tathagatas*, saints and high lamas.

10. Historically, the symbolism and concept of *chortens* can be related to the Buddhist stupas found in India and Nepal.

11. Psychic forces, spirituality and all elements of nature, i.e land, water, air, fire, the solar system and the universe are depicted in the architectural designs of *chortens*.

12. Although, in all there are eight varieties of *chortens*, in the study area three kinds of *chortens* are generally found. All eight types of *chortens* have different structural designs. The eight types are related to enlightenment, heaped lotuses, multiple auspicious doors, great miracle, descent from heaven, reconciliation, victory and nirvana. At Tashiding the *chortens* of enlightenment, reconciliation and great miracle types are created.

The Mahayana Buddhist and Hindu philosophies followed in the study area offer much to ponder on the spiritual and metaphysical nature. The Buddhist relational ethics teach that moral actions are those, which are informed, sensitive to and respectful of all beings. In both, Hinduism and Buddhism, violence to other beings within the biotic community is immoral. The 'eight fold path' of Buddhism is biocentric to the core.

All monumental structures, and the designs and symbols within them are related to biotic life and abiotic environment. Species of the avians wild fauna and domesticated animal life are revered, and find appropriate place in murals and gompa paintings.

The *chortens* rever the lives and works of great saints and high lamas. Since the various elements of nature are depicted in the *chorten* architecture, they teach that we are born from nature, we live in nature, and after death we get rediffused with nature only to be born again as part of the nature.

This thinking has given rise to a practice of deep ecology in the study area. Nature must be continued to be nurtured, and the concept of biocentricism through spirituality must continue to be the driving force with the inhabitants of the region.

Chapter 4 deals with the cultural ecology of the study area, which is pronounced by conservation through traditional ecological knowledge of the local inhabitants. Their life styles have also been affected by isolatory geographical setting, the composition and diffusion of ethnic groups and also the historical inputs. The main findings of the chapter are summarized as follows :-

1. The Upper Rangit Basin where agro-economy and spirituality are more pronounced due to its geographical isolation, was the centre of cultural and political activities till the late eighteenth century, when the capital of Sikkim was shifted from Rabdentse to Tumlong as a consequence of military invasion by Nepal.

2. Through centuries of togetherness, cultural diffusion gathered momentum in almost all disciplines of life including language, methods of praying, hoisting of prayer flags, celebrations of festivals and methodology of food preparation.

3. The present day ethnic culture is pronounced by folklores, shamans and spirit possession. The folklores relating to beliefs, superstitions and customs have been handed over from generation to generation.

4. People of higher income groups and higher levels of education continue to be conservative at heart. The middle class is comparatively less conservative and have accepted Indian and Western cultures along with their virtues and vices. The poorer lot follow their age old traditions.

5. The traditional ecological knowledge of the local inhabitants is qualitative as opposed to quantitive; intuitive; holistic as opposed to reductionsist; moral and spiritualistic as opposed to value-free and mechanistic; is based on centuries old empirical observations as opposed to experimentations; and based on collection of long time information on one locality as opposed to short time-series over large areas.

6. Most rivers, lakes and mountain tops are considered to have spiritual relationships, and therefore considered holy. People believe in divine power in lakes and mountains to release them from distress.

7. Invasive aspect of modern culture, especially use of pollution creating use of polythene, and spread of vegetable dehydrated oils have slowly infiltrated into simple minds of the local folk.

8. All ethnic group viz Limbus Sunuwars, Rais, Tamangs, Bhutias, Sherpas, Kamis and Lepchas have their respective shamanic practices. Most Hindu shamans belong to lower economical and caste status. The shamans are part time professionals as opposed to religious priests. Folk tales on shamans abound.

The practice of shamanism is understandable to the extent of faith healing. Beyond these, people need to be educated towards health, hygiene and faith in modern medical treatment practices.

Knowledge pertaining to traditional ecological know-how must be encouraged, as it helps in conservation of all the natural resources and forests.

The Upper Rangit Basin is a nature's paradise. Festivals such as Bum-chu and others are of a great importance in social life. But the fact that pilgrims attending the festivals cause acceleration of ecological degradation needs to be checked firmly through educational and physical controls.

Afforestation of degraded areas needs immediate attention. Disturbance of avian and land fauna needs to be checked through education and awareness programmes in the region.

From the point of view of human ecology a symbiotic relationship between ethnic groups and religious groups is found positive, and typical of itself in the study area. It is interesting to note that there is an interrelationship between the local inhabitants and their Gods and Goddesses in nature. The visitors to this mountaineous enclave also need to be educated on the value based systems of this local culture, so that invasive degeneration does not seep in.

Chapter 5 presents a comprehensive account of applied ethnobotany of the region. The chapter focuses on the availability and uses of plant life, and the intensity of interrelationship between botanical life and the humans. Summary of findings and the discussion on various ethnobotanical related aspects is given below. The findings could be summarized as follows :-

1. The fertile mountainous enclave, in the study area is a rich source of a variety of plant species. The home gardens, folk markets, various agro

produces, dietary habits of the people and non-timber use of forest products all in unison give an expression of applied ethnobotany in the region.

2. Home gardens are common to almost entire rural habitat within the study area. Vegetables, tubers, fruits bamboo varieties, fodder grass and fodder trees are the main constituents of the home gardens. Varieties of all kharif, rabi and off-season vegetables are grown. Variety of factors affecting diversity include financial position of owners, ground size, configuration and gradient, soil quality and the market prices of the commodities.

3. The home gardens serve both, socio-economic as well as social aspects. Socioeconomic aspects include functions, role of women, cultural ethos and the calendar of activities. Ecological aspects pertain to indigenous knowledge, plant community structure, agro ecosystems, indicator species, and productivity and population interactions.

4. Local weekly folk markets display a commodities diversity of a variety of plants, animals and other products of biological origin. Plant products are either harvested from home gardens or collected from wild. Most vendors are women. Men do the species collection work from the forests.

5. Large variety of indigenous fermented foods and beverages are prepared and consumed in Sikkim. These include kinema, gundruk, sinki, selroti, chhurpi, mesu etc. Marchaa, chhang and raksi are common fermented beverages. Original Tibetans living in higher reaches in Dzongri and Tsoka consume items like momo, thukpa, phing etc.

6. 'Minor forest produce' division of forest department operates five trial plots of medicinal plants in the study area. Variety of species of the medicinal plants grow in the wild, especially around Yuksom. Nepali ethnic practices on medicinal plants pertains to 'jari-buti' system. The Nepali system is deeply seated in antiquity, but is very thinly documented.

7. Spirituality is the mainstay of Tibetan medicine. Tibetan medicines use a mix of plant recipes for prescription, which is either taken in powder form or small rounded tablet form.

8. Non-timber forest produce finds its uses for people's requirements of food, fuel, fibre and medicines. The major species available may be classified under the heads viz bamboos, fodder grasses, fruits, decoratives, mushrooms, yams and fibres.

The home gardens and the ethnic folk markets act as subsistence economy for the weaker sections amongst farmers. Education, awareness and better transportational facilities will help fetch the home garden produce within manageable time to the local markets.

Research work on seed-genetics, and plants stratification in home gardens will help better mapping, planning and production.

Indigenous maintenance of the wild resources of vegetation must be encouraged, as it helps in 'conservation'. By identifying ecologically adaptive aspects of resource management, ethnobotanists contribute to the development of resource-conservative, sustained-yield agro-ecosystems.

As regards medical plants, the aboriginal knowledge is finding a useful mix with modern medicines. This inter-change of 'classical old' with 'scientific new' will help in rescuing the traditional knowledge, improve utility in plant therapy, find new molecular models in plants, and help in obtaining intermediate chemicals.

Chapter 6 deals with the human ecological relationship between the local inhabitants, tourists and the environment. The study area is the gateway to the summit of Mount Khanchendzonga, the base for which is located at Yuksom. The nature trail from Yuksom to Dzongri is one of the most popular hiking treks for mountain adventurists. The monasteries and holy lakes are the other pull factors and attraction to the tourists. The main findings of the chapter are summarized below :-

1. Tourists arriving in Sikkim are generally of three categories. Mountaineering and adventurers who are lured by the call of the mountain peaks and glaciers. The sight seeing types who come to feel the clean pollution free atmosphere and fascinating mountain views. The third category belongs to spirituality seekers, who visit the monasteries, gompas and other historical sights.

2. Indians from the plains, and foreigners from western countries visit the study area. A large number of foreign tourists traverse the basin on motor cycles, bicycles, and some of them through self contained trekking. HMI Darjeeling conducts mountaineering training, the camp sites for which are located at Yuksom and Dzongri.

3. The Khanchendzonga National Park the Mainam Wild Life Sanctuary form the protected areas, and hence permission for entry is granted through the office of Chief Wildlife Warden.

4. Unique fauna and flora exists in the study area. The Western watershed of Rathongchu contains vast mountain slopes and spurs covered with varieties of rhododendrons. The floristic biodiversity lands itself to detailed ethnobotanical research, thereby motivating research groups and individuals to visit and survey the area.

5. The Upper Rangit Basin being the original seat and capital centre of the earlier Chogyals, has a rich history. The historical ruins at Rabdentse, and Dubdi are attractions.

6. Some of the oldest Buddhist monasteries are located in the Basin, which include Dubdi, Pemayangtse, Tashiding and Ralang. There is a Bon monastery and, one monastery of the Gelugpa sect located at Ravangla. Apart from being places of religious discourses, worship and meditation, they are store houses of religious books and murals. Spiritual tourists come for casual visits as well as for the purposes of research of spirituality.

7. Tourism impacts in the region pertain to both, positive and negative aspects. These impacts pertain to environment, economy, socio-cultural and human ecological fields.

8. Litter, garbage and pollution along the trails and at camping sites are major negative impacts on environment. Independent trekkers and local people are main sources of littering. The demand of fuel for firewood results in deforestation along the trails.

9. Economically, a large section of local inhabitants benefit from the tourist inflow. These include hoteliers, transporters, handicrafts centres, the local guides and the porters.

10. As regards human ecology, the interaction with tourists bring about changes in behaviour, dress, life styles and social culture, some times at the cost of losing out on the values of traditional culture. The positive affects include poverty alleviation, awareness generation, infrastructure development, research and publicity.

11. The government, and the local officialdom need to understand the definition of 'Sustainable mountain tourism and ensure functioning of all tourism related institutions within those norms.

Economically, it should seek to maximize human welfare within the constraints of existing capital stock and technologies.

From the points of view of environment, the carrying capacity of the eco-system needs to be taken into account. The natural resources must not be disturbed.

Human ecology is a key factor. Social organizations are crucial to ensure proper symbiotic interrelationships, without prejudice to preservation of local traditional cultures.

Chapter 7 presents the sum total of human ecological aspects of the region and also highlights the local psyche in the context of the indigenous cultural set up. The main observations are as follows :-

1. The social structure of inhabitants is based on social status ascribed mainly by tribal and non-tribal affiliations. The Nepali community therefore has a feeling that the Bhutia and Lepcha communities, though economically better off, yet enjoy more constitutional liberties.

2. Invasion of modernity and popular culture has triggered intra-

contradictions within the society, inter-contradictions of the ethnic groups, and contradictions with the popular culture itself. Hence mutual incompatibility between traditional modernity may be witnessed in a few sections of the youth.

3. The isolatory setting has forced secludedness of the inhabitants. This has helped in diffusion between various communities, and preservation of traditional culture and ecological wisdom. However, this has also denied opportunities of higher education to the youth.

4. Democratisation of the state and its merger with India has left certain people confused with regards to their identity. The liberties of democracy have not been fully understood, giving rise to easy life, and efforts towards making easy money.

5. The belief and faith in the Lamas, and Shamans is almost absolute. Every hazard, natural or physical is related to spirituality and the writ of the monks on their 'pronouncements' is taken to be gospel.

6. Sikkim being a less densely populated state, the level of competition is low. The inhabitants as a result have become complacent in their attitude towards hard work in the fields of education beyond senior secondary classes.

7. Personal discipline of the inhabitants is of a high order, and the law and order problems are almost non-existent.

8. Non-violence is the way of life of the inhabitants, and knowingly they do not kill or harm the plant and animal life.

The personal traits of the inhabitants clearly indicate them to be God-fearing, independent natured well disciplined, with absolute faith in spirituality and the religious heads. They continue to be highly superstitious.

Knowledge, awareness and education in their case must therefore relate to :-

(a) Not to lose 'what they know'

(b) Learn 'new' that is relevant and necessary.

A spirit of competitiveness in field of education is required, which can only come through political decision of earmarking a certain percentage of marks to be obtained as a prerequisite for seeking jobs.

As a people, they must be identified by the mainstream country-men as a part of the nation. Slowly, the Sikkimese will understand the meaning of state identity and national identity, thus preserving their indigenous Sikkimese culture as nationals of India.

SECTION II
SUGGESTIONS AND RECOMMENDATIONS

There is no doubt that future is, and always has been, one of the most motivating subjects for thought, planning and implementing action oriented programmes. The present day conditions have come about as a result of human actions within their respective human and natural ecosystems in the past. Health of future ecosystems of all kinds will depend upon the actions of the present generation.

So as to ensure healthy interactions and better prospects for future on a long term basis, the suggestions and recommendations require a holistic view and understanding. Reductionist views and action plans may result in multiplicity of efforts in a few fields, at the cost of other elements pertaining to human ecology. Keeping the summary of findings in view, a set of suggestions and recommendations is listed below.

1. Sikkim has a rich spiritual and culture ethos. These policy paradigms of the state government must include the conservations of these human ecology attributes in all the action oriented programmes for development. These must include :-

(a) Incorporate human ecological ethics into all academic and institutional practices.

(b) Encourage environmental responsibility amongst all ministries, government bodies and departments

(c) Maintain updated audits of the environmental contents of all public and private sector industries.

(d) Promote energy conservation and the re-use and recycle of all possible materials at individual, village, institutions, district and state levels.

2. The planning policy must judiciously intermix the use of modern scientific techniques with the traditional ecological knowledge of the local inhabitants.

3. The existing roads and tracks in the Upper Rangit Basin need a constant maintenance, as some of them pass through geologically unstable stretches. Constructions of new roads require a systematic ground survey. The SPWD (Road and Bridges) is recommended to be modernized in terms of equipment and training in this regard.

4. Constructions of bridges, dams dwellings and also terrace farming in some areas have resulted in fragmenting and fracturing of a few rock faces and re-entrants. It is recommended that watershed treatment plans be taken up on priority for all such locations. The rock face on road to Khecheopalri Lake and Yuksom are in need of urgent treatment.

5. There are some artisans, who are well trained in the skill of building and maintaining monasteries, gompas and *Chortens*. There are others, who carry out intricate painting and designing work. It is recommended that regular educational and training institutions be opened for such artisans and painters. Although handicraft centres do exit at some places like Kewzing, the training in these centres requires to be augmented further. Trained architects are recommended to be employed in the institutes, who should study the amalgamation of traditional designs with modern techniques.

6. Spiritual teachings amongst the Buddhist, Hindus and other minority groups should include the religious relevance to human ecology and deep ecology. The traditional *Karma* should incorporate ecological *Karma* within its folds. The recommended models of *Karma* are represented as under :-

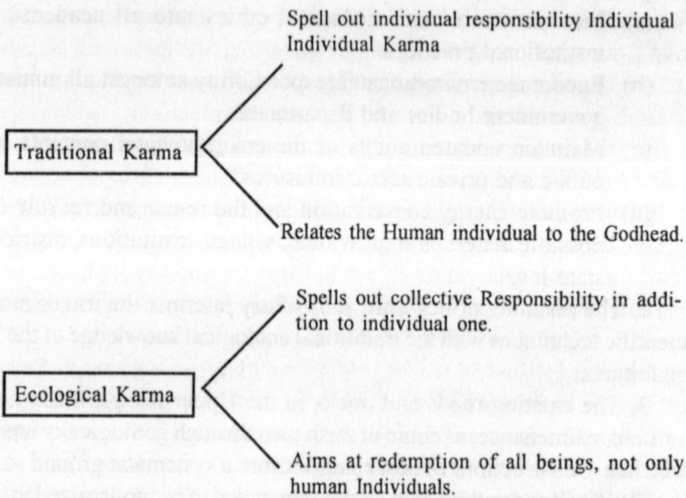

7. The intellectuals, government officials and religious heads should seek a healthy inter-mix of 'traditional ecological knowledge' with modern scientific techniques. While the constructive aspects of 'nature conservation' must be further cemented, the invasive ideas of popular culture which attract youth towards 'drug addiction' etc need to be

guarded against. The conservation of all holy places such as Khelcheopalri Lake, the monasteries, the fauna and flora, and the general habitat is recommended to be done through traditional ecological knowledge of the inhabitants.

8. There is need recognized for training workshops to develop ethnobotany for application in the fields of :-

(a) Management and conservation of plant resources.

(b) Capacity building in terms of knowledge and materials.

(c) Commencement of action- oriented field projects in Upper Rangit Basin.

9. At the government and organizational levels there is a need to study and review the range of methods available for optimum output and sustainable benefits from the home gardens to include species inventory, frequency and composition. Structures of home gardens composition, and the architectural profile of the home gardens are recommended to be researched further by trained ethnobotanists for the benefits of local inhabitants.

10. It is recommended that a computer programme be prepared for plant identification for 'Non –timber forest produce' uses, and also for medicinal plants. Such an inventory will be beneficial to students, field workers and also the local inhabitants.

11. Following broad areas are recommended for development of community based ecotourism strategy :-

(a) Mechanism for local communities to directly benefit from ecotourism revenues.

(b) The development of financial and legal mechanisms.

(c) Training of local inhabitants in garbage management, trail repairs, plantation work and management, conservation education and the aspects of lodge operating, cooking and acting as guides.

12. The entire area of Upper Rangit Basin or for that matter Sikkim is recommended to be brought under the 'Man and Biosphere' (MAB) programme of UNESCO, and be declared as a 'biosphere reserve'. This will ensure protection of the ecosystems and biodiversity within the state. This will also help in research and monitoring of the habitat through UN agencies. The UNESCO will by itself take up the responsibility of education, awareness and upliftment of the local inhabitants, while conserving their cultural heritage.

Bibliography

BOOKS

1. Agarwal, A; Narain S., and Sen Srabani (eds) *The Citizens (Fifth Report), Part I :* National overview, Center for Science and Environment, New Delhi, 1999.
2. _______________, *The Citizens (Fifth Report), Part II : Statistical Database*, Centre for Science and Environment, New Delhi, 1999.
3. Agrawal, M., M., (ed), *Ethnicity, Culture, and Nationalism in North-East India,* Indus Publishing Company, New Delhi, 1996.
4. Ali, A. (ed), *Environmental Protection of the Himalaya : A Mountaineers View,* Indus Publishing Company, New Delhi, 1994.
5. Barua, M., *Habitat and Economy in an Eastern Himalayan Enclave: Case Study of the Rangit Basin.* University of Calcutta, 1988.
6. Basnet, L., B., *Sikkim : A Short Political History,* S Chand & Co. Pvt. Ltd, New Delhi, 1974.
7. Bennett, J.W, The *Ecological Transition : Cultural Anthropology and Human Adaptation*, Pergamon Press Inc., 1976.
8. Bhanja, K., C., *History of Darjeeling and the Sikkim Himalaya,* Oxford Book and Stationery Co. Darjeeling, 1993.
9. Bhasin, M.K., and Bhasin, V., *Sikkim Himalayas Ecology and Resource Development*, Kamla Raj Enterprises, Delhi, 1995.
10. Bhasin, Veena, *Ecology, Culture and Change,* Mittal, M.C., Inter-India Publications, New Delhi 1989.
11. Bhat, H,P., *Environmental Dimensions of Rural Settlements In Hill Areas*, Ashish Publishing House, New Delhi, 1993.
12. Bhatt, K., N., *Uttarakhand : Ecology, Economy and Society*, Horizon Publishers, Allahabad, 1997.
13. Bhattacharya, R., N., *Development and Environment : A Perspective, for Hill Regions, In, Problems and Strategies of Development In the Easter Himalaya,* Dhamala, R., R., (ed), Gyan Publishing House, New Delhi, 1993.
14. Bhattacharya, P., K., *Aspect of Cultural History of Sikkim : Studies In Coinage,* K.P. Bagchi & Company, Calcutta, 1984.

15. Biswas, S., *Development Strategy in the Hill areas of Darjeeling : A Critical Appraisal*, in, Problems and Strategies of Development in the Eastern Himalaya, Dhamala R.,R., (ed), Gyan Publishing House, New Delhi, 1993.
16. Blij, H., J., and Muller, P.,O., *Physical Geography of the Global Environment,* John Wiley & Sons, New York, 1993.
17. Bookchin, M., *The Philosophy of Social Ecology*, Rawat Publications, Jaipur, 1996.
18. Boot, D.,P., *Monasteries In Sikkim : A Geographical Study*, Centre For Himalayan Studies, NBU, Siliguri, 1996.
19. Brazier, D., *The Feeling Buddha : A Buddhist Psychology of Character, Adversity and Passion*, Harper Collins Publishers, India, 1977.
20. Burbank, J., *Cultural Shock! Nepal,* Times Books International, Singapore, 1992.
21. Burman, B., K., R., *Basic Approach to Planning and Development In North-East India*, In, Problems and Strategies of Development In the Eastern Himalaya, Dhamala, R.,R., (ed), Gyan Publishing House, New Delhi 1993.
22. Burne, C.,S., The Handbook of Folklore, Sidgwick & Jackson Ltd, London, 1914.
23. Biswas, K.,P., *Plants of Darjeeling and Sikkim Himalayas : Vol. I* West Bengal Govt. Press, Alipur, 1967.
24. __________, *Medicinal Plants of Darjeeling and the Sikkim Himalayas*, West Bengal Govt. Press, Alipur, 1956.
25. Blache, V.de L; *Principles of Human Geography*, London, 1950.
26. Bruce, C.,G., *Twnety Years in the Himalaya*; Arnold ; London ; 1910.
27. Burrard, S.,G., & Hayden, H.H.A.; *A Sketch of the Geography and Geology of the Himalaya Mountains of Tibet*; Revised by Sir Sidney Burrard and A.M. Heron; Delhi; 1933.
28. Chakravarty, P.,K., *Problems and Strategies of Development in the Darjeeling Himalaya,* In, Problems and strategies of Development in the Eastern Himalaya, Dhamala, R.,R., (ed), Gyan Publishing House New Delhi, 1993.
29. Chopra, P.,N., *Sikkim*, S.Chand & Company Ltd, New Delhi, 1979.
30. Chaudhary, N.,C., and Sarkar, K, *Hill Villages of Darjeeling : Aspects of Economy, Ecology and Agricultural Modernization*, In, problems and strategies of Development in the Eastern Himalaya, Dhamala, R.,R.,(ed), Gyan Publishing House, New Delhi, 1993.
31. Chaube, S., K., (ed), *The Himalayas; Profiles of Modernistion and Adaptations*; Sterling Publishers Private Limited, New Delhi, 1995.
32. Chorley, R.,J., and Haggett, P., *Models in Geography,* Methuen & Co, London; 1977.

33. Coelho, V.,H., *Sikkim and Bhutan,* Indian Council for Cultural Relations; New Delhi; 1970.
34. Dalai Lama, H.H; The, *The Four Noble Truths.* Harper Collins Publishers, India, 1997.
35. Das, B.S., *Sikkim's Identity as an Indian State,* In, Sikkim : Society, Polity, Economy, Environment, Mahendra, P. Lama,(ed), Indus Publishing Company, New Delhi, 1994.
36. __________, *The Sikkim Saga,* Vikas Publishing House Pvt., Ltd., New Delhi, 1983.
37. Datta, A., *Sikkim Since Independence : A study of Impact of Education & Emerging Class Structure,* Mittal Publications, 1991.
38. Dutta, K., *Inter-Ethnic Relations In Sikkim In Historical Perspective,* Occasional paper No 11.NBU., Siliguri, 1997.
39. Day, D., *The Eco Wars: A layman's Guide to the Ecology Movement,* Harpar Books Ltd., London, 1989.
40. Delattre, P., *Tales of a Dalai Lama* Victor Gollancz Ltd., London, 1972.
41. Dhamala,R.,R., *Panchayati Raj Institutions In Sikkim as Instruments of Development,* In, problems and strategies of Development in the Eastern Himalaya, Dhamala, R., R.,(ed), Gyan Publishing House, New Delhi,1993.
42. __________, *Monasteries of Sikkim*, Occasional Paper-8, University of North Bengal, Siliguri, 1993.
43. Dorjee,P., *Healing Stress, Anxiety and Depression through Tibetan Medicine*, Kailash Centre of Oriental Medicine, London, 1998.
44. Das, S., C., *Indian Pundits in the Land of Snow,* Baptist, Mission Press; Calcutta, 1893.
45. Detwylec, T.,R., *Man's Impact on Environment,* McGrow Hill Book Company, 1971.
46. Donaldson,F., *Lepcha land or Six Weeks in the Sikkim Himalayas,* Simpson-low-Marston; London, 1900.
47. Ember, C.,R., Ember,M., *Anthropology,* Prentice Hall of India Private Limited, 1993.
48. Fantin, M., *Sherpa-Himalaya-Nepal,* Toppan Co., Pvt., Ltd., Singapore, 1976.
49. Fischer,F., and Hajer, M.A.,(eds), *Living with Nature : Environmental Politics as Cultural Discourse,* Oxford University Press, New York,1999.
50. Foning,A.,R., *Lepcha : My Vanishing Tribe,* Sterling Publishers Private Limited, New Delhi, 1987.
51. Freshfield, D.,W., *Round Kanchendjunga : A Narrative of Mountain Travel and Exploration,* First published in 1903, Reprinted, Ratna Pustok Bhandar, Kathmandu, 1979.

52. Furer-Haimendorf,C., Von, ed, *Caste of Kin in Nepal, India and Ceylon, Anthropological Studies in Hindu-Buddhist Contact Zones*, Asia Publishing House, Bombay, 1966.
53. ____________, *Asian Highland Societies in Anthropological Perspective,* Sterling Publishers, New Delhi, 1981.
54. Glaeser, B., *Environment, Development Agriculture : Integrated Policy through Human Ecology*, UCL Press, London, 1995.
55. Goldstein, M.C., Beall, C.M., *Nomads of Western Tibet : The Survival: A way of Life,* Serindia Publications, London, 1989.
56. Govinda, L.,A., *Foundations of Tibetan Mysticism,* B.,I., Publications Pvt.Ltd., New Delhi, 1977.
57. Gupta, M.,D., *Problems of Economic Development in Bhutan,* In, Problems and strategies of Development in the Eastern Himalaya, Dhamala, R., R., (ed), Gyan Publishing House, New Delhi, 1993.
58. ____________, *Sikkim : Problems & Prospects of Development,* Indus Publishing Company, New Delhi, 1992.
59. Gupta,R.,(ed), *The Mis-management of Disasters,* IDNDR, Geneva, 1999.
60. Gorer, G., *The Lepchas of Sikkim;* First Published as the "Himalayan Village, An account of the Lepchas of Sikkim" by Basic Books, 1967. Reprinted by Cultural Publishing House, Delhi- 1984.
61. Grover, B.S.K., *Sikkim and India, Storm and Consolidation,* Jain Brothers, NewDelhi, 1974.
62. Held, S., *The Monasteries of the Himalayas : Tibet, Bhutan, Ladakh, Sikkim*, ISBN, Italy, 1987.
63. Hitchcock, J., T., and Jones, R., L., (eds) *Spirit Possession in the Nepal Himalayas,* Vikas Publishing House Pvt. Ltd., New Delhi, 1994.
64. Honigmann, J., J., (ed) *Handbook of Social and Cultural Anthropology (Vol-1),* Rawat Publications, Jaipur and New Delhi, 1997.
65. ______________,(ed), *Handbook of Social and Cultural Anthropolgy (Vol-2),* Rawat Publications Jaipur and New Delhi, 1997.
66. Husain, S., A., *The National Culture of India*, National Book Trust, India, 1978.
67. Hooker, J., D., *The Rhododendrons of Sikkim Himalaya,* London, 1849.
68. ____________, *Illustration of Himalyan Plants*, London 1855.
69. ____________, *Himalayan Journals, Notes of a Naturalist in Bengal , the Sikkim and Nepal Himalayas, the Khasa-Mountans, &C, 2 Vols,* Murray, London, 1855.

70. __________, *The Flora of British India,* London, 1873-1897.
71. __________, *A Sketch of the Flora of British India,* London 1904.
72. Inglis, J., T., (ed) *Concepts and Cases : Traditional Ecological Knowledge*, IDRC, 1993.
73. Ives, J., D., & Messeri, B., *The Himalayan Dilemma,* UN University Routledge, London, 1989.
74. Jha, S., K., and Mishra, S., N., *Sikkim : Government & Politics,* Ess Ess Routledge, London, 1989.
75. Jhaveri, S., (ed) *Sikkim : Our World in Colour,* The Guidebook Company Limited, Hong Kong, 1990.
76. Jina, P., S.,(ed) *Recent Rescarches on the Himalaya,* Indus Publishing Company, New Delhi, 1997.
77. Jordan, T., G., & Rowntree, L., *The Human Mosaic : A Thematic Introduction to Cultural Geography*, Harper Colling Publishers, 1990.
78. Kandell, A., S., and Salisbury, C., Y., *Mountaintop Kingdom, Sikkim,* Vikash Publishing House Pvt Ltd., New Delhi- 1972.
79. Kar, R., K., *A Small Tribe in the Eastern Himalaya, An appraisal of its problems and Development,* in, Problems and Strategies of Development in the Eastern Himalaya, Dhamala, R., R., (ed), Gyan Publishing House, New Delhi, 1993.
80. Karan, P., P., and Jenkins, Jr. W., M., *The Himalaya Kingdoms, Bhutam, Sikkim and Nepal,* D.Van Nostrand Co. Inc. New York, 1963.
81. Kazi, J., N., *Inside Sikkim : Against the Tide*, Hill Media Publications, Gangtok, 1993.
82. Khati, D., (ed) *Sikkim Tourism, Sikkim Paradise Unlimited,* Sikkim Information update Sikkim Telegraph Publication House, 1997.
83. Kiyota, M., (ed) *Mahayana Buddhist Meditation : Theory and Practice*, Motilal Banarsidass Publishers, Delhi, 1978.
84. Kotturan, G., *Folk Tales of Sikkim,* Sterling Publishers Private Limited, New Delhi, 1989.
85. __________, *The Himalayan Gateway : History and Culture of Sikkim*, Sterling Publishers Pvt. Ltd. New Delhi 1983.
86. Krishnamurti, J., *This Matter of Culture,* Krishnamurti Foundation, India, Madras, 1992.
87. Kuhn, Delia. and Ferdinand; Ed, *Edge of Tibet, Sikkim and its Neighbours in their Borderlands,* Knoph, New York, 1962.
88. Lal, J., S., (ed) *The Himalaya : Aspects of the Change,* Oxford University Press, 1981.
89. Mahajan, B., Majumdar, S& Agnihorti, D., C., *Educational Administration in Sikkim Structures, Processes and Future Prospects*, Vikas Publishing House Pvt. Ltd. 1995.

90. Mann, R., S., *Aspects of Indian Social Anthropology,* Concept publishing Company New Delhi, 1996.
91. Merchant, C., (ed) *Ecology : Key Concepts in Critical Theory,* Rawat Publications, Jaipur 1996.
92. Misra, B., P., *Economic Development of Bhutan Problems & Strategies*, in, Problems and stregies of development in the Eastern Himalaya, Dhamala, R., R., (ed), Gyan Publishing House, New Delhi, 1993.
93. Morris, J., *Living with Lepchas,* A Book about the Sikkim Himalayas, Heinemann, London, 1938.
94. Mridula, Datt, N., *Ecology and Tourism*, Universal Publishers Distributors, Delhi, 1991.
95. Murthy, M., S., R., and Lakshmi, V., U., *Catchment Characterisation and area Speeding Treatment Plans*, Rangit Hydel Power Project, Sikkim, National Remote Sensing Agency, Hyderabad.
96. Narayana Swamy, K., S., Krishna Murthy, T., (eds), *Sustainable Development : The Gandhian Perspectives,* Birth Centenary Committee, Karnataka, 1993.
97. Narayana, V., V., D., Sastry, G., and Patnaik, U., S., *Watershed Management*, ICAR, 1990.
98. Nepali, G., S., *Some Sociological Aspects of Development in Eastern Himalaya*, In, Problems and strategies of Development in the Eastern Himalaya, Dhamala, R., R., (ed) Gyan Publishing House, New Delhi, 1993.
99. Norbu, D., *Changes In Tibetan Economy, 1959-76*, in.problems and strategies of Development in the Eastern Himalaya, Dhamala, R., R., (ed),Gyan publishing House, New Delhi, 1993.
100. Nebresky-Wojkowitz, R.Von, *Where the Gods are Mountains, Three years among the people of Himalaya. Three years among the people of Himalayas*. (Translated from German by Michael Bullock) Reynal, New York, 1957.
101. Olschak, B., C., Gansser, A., and Buhrer, E., *Himalayas : growing Mountains, Living, Myths, Migrating People*, Bookwise, *New Delhi, 1988.*
102. Odum, E., P., *Fundamentals of Ecology,* Saunders College Publishing, Philadelphia, 1971.
103. Pasayat, C., *Tribe, Caste and Folk Culture*, Rawat Publications, Jaipur, 1988.
104. Pelto, G., H., Pelto, P., J., and Messer, E., (ed), *Research Methods in Nuritional Anthropology*, The United Nations University, 1989.
105. Pradhan, U., C., Lachungpa, S., T., *Sikkim-Himalaya Rhododendrons,* Primulaceae Books, West Bengal, India, 1990.

106. Prime, R., *Hinduism and Ecology : Seeds of Truth*, Motilal Banarsidass Publishers, Delhi, 1994.
107. Pallis Marco, *Peaks and Lamas*, Cassel and Company Limited, London, 1939.
108. Pares, Bip., *Himalayan Honeymoon*, Hodder & Stoughton, London, 1939.
109. Rai, L., and Sharma, E., *Medicinal Plants of the Sikkim Himalaya : Status, Usage and Potential*, Bishen Singh, Mahendra Pal Singh, Dehradun, 1994.
110. Rai, S., C., Sundriyal, R., C., and Sharma, E., (eds) *Sikkim : Perspectives for Planning and Development*, Bishen Singh Mahendra Pal Singh, Dehradun, 1998.
111. Rai, T., Rai, L., *Trees of the Sikkim Himalaya.* Indus Publishing Company, New Delhi, 1994.
112. Rastogi, A., Godbole, A., and Shengji. (eds), *Applied Ethnobotany in Natural Resource Management- Traditional Home Gardens.* International Centre for Integrated Mountain Development, Nepal, 1998.
113. Rustomji, N., *Sikkim : A Himalayan Tragedy,* Allied Publishers Private Limited 1987.
114. __________, *Enchanted Frontiers, Sikkim, Bhutan and Indian's North-Eastern Borderlands*, Oxford University press, Calcutta, 1971.
115. Risley, H,H., (ed) *The Gazetteer of Sikkim*, Bengal Govt Secretariat, Calcutta, 1894.
116. __________, *The Tribes and Castes of Bengal , Ethonographic Glossary, 2 Vols* Bengal Secretariat Press, Calcutta, 1892.
117. Raha, M K, *The Himalayan Heritage*, Gian Publishing House, 1987.
118. Rao, P.R., *India and Sikkim 1814-1970*, Sterling Publishers (p) Ltd. 1972.
119. Saklani, A., & Jain, S., S. *Cross-Cultural Ethnobotany of North East India*, Deep Publications, New Delhi, 1994.
120. Sakya, J., B., (Compiler), *Short Discription of Gods, Goddesses, and Ritual Objects of Buddhism and Hinduism in Nepal*, Handicraft Association of Nepal, 1989.
121. Sangharakshita, B., *The Three Jewels: An Introduction to Buddhism,* Rider & Company London, 1967.
122. Schultes, R,E., and Reis, S., (eds), Ethnobotony : Evolution of a Discipline, Chapman & Hall, London, 1995.
123. Seeland, K., and Schmithilsen, F., (eds), *Man in the Forest*, D., K., Printworld(P) Ltd., New Delhi, 2000.
124. Sengupta, N., *State Government and Politics : Sikkim,* Sterling Publishers, Private Limited, New Delhi, 2000.

125. Simpson, W., *The Buddhist Praying –Wheel*, Asia Educational Services, New Delhi, 1891.
126. Singh K., S., (ed) *People of India : Sikkim (Vol XXXIX)*, Seagull Books, Calcutta, 1993.
127. Singh, R., Y., *Geography of Settlements,* Rawat Publications, Jaipur, 1994.
128. Skolimowski, H., *Dharma, Ecology & Wisdom in the Third Millennium*, Concept Publishing Company, New Delhi,1999.
129. Stocks, C., D., B., *Sikkim : Customs and Flok-lore*, Cosmos Publications, Delhi, 1975.
130. Subba, J., R., *Agriculture in the Hills of Sikkim,* Sikkim Science Society, Gangtok , 1984.
131. ____________, *TheLimboos of the Eastern Himalayas : With Special Reference to Sikkim,* SYMS, Gangtok, 1999.
132. Subba, T., B., *Dynamics of a Hill Society,* Mittal Publications, Delhi, 1989.
133. ____________, *Politics of Culture- A Study of Three Kirata Communities in Eastern Himalayas*, Orient Longman Ltd, Chennai, 1999.
134. Subba, T., B., and Datta, K., *Religion and Society in the Himalayas,* Gian Publishing House, New Delhi.
135. Schappert, L., G., *Sikkim, 1800-1968*, An Annotated Bibliography, East West Centre, Hawali, 1968.
136. Sinha, A., C., *Politics of Sikkim*, A Sociological Study, Thomson(India), Delhi, 1975.
137. ____________, *Studies in the Himalayan Communities (Text& Context)*, Book Today (Oridental Publishers), New Delhi, 1983.
138. Sinha, O., P., *Strategic Sikkim*, B, R, Publishing Corporation, Delhi, 1985.
139. Tamsang, K., P., *The Unknown and Untold Reality about the Lepchas*, Lyangson Tamsang, Hongkong, 1983.
140. Tashi, T., *Fundamentals of Tibetan Medicine*, Men-Tsee-Khang, New Delhi, 1981.
141. *Tibet : Environment and Development Issues,* Department of Information and International Relations. Central Tibetan Administration Dharamsala, India, 1992.
142. Tandulkar, G., M., *Kiratese at a Glance*, International Standard Books and Periodicals Store, Nepal, 1980.
143. Thakur, R., N., *Himalyan Lepchas,* Archives Publishers, Distributors, New Delhi, 1988.
144. Thurman, R., A., F., *Essential Tiberan Buddhism,* Harper Collins Publishers India, 1995.

145. Toffler, A., *The Third Wave,* William Collins Sons & Co. Ltd. London, 1980.

146. Toynbee, A., J., *A Study of History, Vol.1* Dell Publishing Co., New York, 1965.

147. Verma, R., *Sikkim : A Guide and Handbook,* S.Verma, Gangtok, 1990.

148. Vidyarthi, L., P., & Jha, M., (eds) *Ecology, Economy and Religion of Himalayas*, Orient Publications, Delhi, 1986.

149. Waddell, L., A., *Among the Himalayas,* Ratna Pustok Bhandar, Kathmandu, Nepal, 1899.

150. White, J. Claude, *Sikkim & Bhutan : Twenty-one Years on the North-East Frontiers,* Manas Publications Delhi, 1909.

151. Wilson, B., Ikeda, D., *Human Values in a Changing World*, Lyle stuert Inc., Secaucus, NJ, 1987.

152. Wittman, J., S., Jr., (ed) *Selected Articles in Social Ecology,* MSS Information corporation, New York, 1974.

153. Ware, J., E., *Report on a visit to Sikkim and the Tibetan Frontier in October, November and December, 1873*, Bengal Secretariat Press, Calcutta 1874.

154. Yeshe, L., & Rinpoche, Z., *Wisdom Energy : Basic Buddhist Teachings*, Wisdom Publications, 1976.

155. Younghusband, Francis, *The Heart of Nature or the Quest for Natural Beauty*. Murray, London, 1921.

156. Zetland, L.,J., L.,D., *2nd Marquis of, Lands of the Thunderbolt, Sikkim, Chumbi and Bhutan*, Constable, London, 1923.

ARTICLES

1. Basu, S., "Bhots of North Sikkim", *Man in India*, Vol. 46, No 2. April-Jun3, P. 164-171, 1966.

2. Buchanon, W., J., "In the Foorsteps of Hooker Through Sikkim and Nepal, A Centenary Tibute", *Bengal Past and Present*, XIV, P 252-274, Calcutta, 1917.

3. Benerjee, H and Chakraborti B. On the lesser Himalayan Granite Gnesses of the Darjeeling-Sikkim Area. Special publication *Geological Survey of India*, 12, 1984, P. 251-259.

4. Bhutia Kunga Gyatso. *Crop Statistics of Sikkim*, 1990-1992, Department of Agriculture, Government of Sikkim Gangtok, 1992.

5. Bhutia D.T., Gupta R., K., Biswas A., K., *Soil Bulletin on Fertility Status of the Soils of Sikkim*, Department of Agriculture, Government of Sikkim, Gangtok, 1986.

6. Burkill, I, HY Alpine Notes from Sikkim, *New Bulletin* 1907.P.92-94.

7. Chandra, P.R., Sen C. R. A New Carboniferous Trilobite from Sikkim Himalaya. *Geological Survery India*, Miscellaneous Publ.41(2), 1976.P.64-96.

8. Campbell, Archibald, "Note on the Lepchas of Sikkim with a vocabulary of their Language", *Journal of the Asiatic Society of Bengal*, IX, Part I.P. 379-390, 1840.

9. __________, "Journal of a trip of Sikkim, in December, 1848, with a sketch Map", Journal of the Asiatic Society of Bengal, XVIII, Part I.P. 482-541, 1849.

10. Dasgupta N., and Mukhopadhoya C. The Land Question of Sikkim. *The Indian Journal of Hill Economy*, Vol.3.December, 1978.

11. Debnath J.C.Evolution of the Sikkimese Economy during the British period. *Indian Journal of Hill Economy*, Vol 2, 1977 March.P. 20-37.

12. Duncan, P.C. Sikkim Rhododendrons, *Himalayan Journal,* 7, 1935.P. 111-112.

13. Gammie G.A. Botanical Exploration in Sikkim, Tibet Frontier, *New Bulletin*, 1893.p.297-315.

14. __________, Report on a Botanical Tour in Sikkim, *Records of Botanical Survey of India*, 1 No 2.P. 1-24,, 1894.

15. __________, Vegetation of Temperate and Alpine Sikkim. In *(the Gazetteer of Sikkim*, Introd, Risley H.H., Calcutta. Bengal Government, 1894, Reprinted by Sikkim Nature Foundation, Gangtok, 1989.P. 86-111).

16. Gregery, J.W., A low level Glaciated surface in the Eastern Himalaya, *The Geological Magazine,* VI.P. 397-406, 1919.

17. Hooker, J,D., On the climate and vegetation of the Temperate and Cold Regions of East Nepal and the Sikkim Himalayan Mountain. *Journal of Horticulture Society of India.* Vol.8, 18654. P. 35-65, 73-95.

18. __________, On the Physical Geography of Sikkim Himalaya. *Hooker Journal of Botany,* Vol.8, 1851.P.21-31.

19. Hope Namgyal The Sikkimese Theory of Land-holding and the Darjeeling Grant. *Bulletin of Tibetology,* Vol. 3(2), 1960, July, P.1.

20. Jeffers, J., N., R., (ed), *The International Journal of Sustainable Development and World Ecology*, Vol.1 No 1, The Parthenon Publishing Group, London, 1994.

21. Karan, P.P., Geographic Regions of the Himalayas, *Bulletin of Tibetology*, III No.2. P. 5-27, 1966.

22. King, Sir George and Pantlling Robert, The Orchids of the Sikkim Himalaya, *Annals of the Royal Botanic Garden, Calcutta,* Vol VIII,P. 1-4, P.348-448.

23. Kumar, S., Some New Records of Grasses from Sikkim., *Journal of Indian Botanical Society*, Vol.72, 1993.P. 165-166.

24. Lama, P. Mahendra., Sikkim Economy and Environment, *Himalaya* Vol.5(2) June 1994 P.42.

25. Lioyd George, Wild dogs of Sikkim, *Sikkim Herald,* Vol.2 (214) December 18, 1970.

26. Majumdar, A.K., Hooker and Campbell in '*Sikkim, Bengal Past and Present*'. LXXXI, P. 132-139, July, 1962.

27. Mani, M.S. The Himalaya, its Ecology and Bio-geography, A Review. In Pangey, Y.P.S. and Rawal, R.S. eds. *High Altitudes of the Himalaya*, Nainital, Gyanodya Prakashan, 1994.

28. N., Nita, Lepchas of Sikkim, Bulletin of Tibetology, New Series, 6,2,P. 18-23, 1982.

29. Paljor S., Prospect of Livestock. Development in Sikkim. Paper Presented at (Workshop on Land Husbandry System in the Himalayan Region, *Sikkim Science Society Gangtok,* April 19-20, 1982.

30. ________________, Animal Production of Sikkim in the year 2000 A.D Paper Presented at (Workshop on Science & Technology inputs for socio-economic development of Sikkim, *Sikkim Science Society, Gangtok*, August 24-25, 1987).

31. Risley, H.H, History of Sikkim, Appendix II of the *Journal of the Buddhist Text Society of India*, IVPO.6-16, 1896.

32. Sherwill, J.L., Journal of a Trip Undertaken of Explore the Glaciers of the Kanchanjingah Group in the Sikkim Himalaya in November, 1861, *Journal of the Asiatic Society of Bengal,* XXXI,P.457-471, 1862.

33. Sinha Roy, S., Alluvial Cones in Relation to the Landscape Morphology of the Rangit Valley Himalaya", Himalayan Seminar, *Geological Survey of India,* Miscl, Pub.Part (2), 24, P. 423-428, 1971.

34. ________________, Terrace Shifts in the Rangit Valley, Eastern Himalayas and their Significance, *Seminar on Geomorphology, Geohydrology and Geotectonics of Lower Ganga Basin*, IIT, Kharagpur, 1972.

35. Schaefer, L.A. Sikkim Awakening. *Himalayan Magazine Vol5(5), 1995. P. 8-16.*

36. Sharma, D.,D and Ghosh, B., Contribution to the flora of Sikkim Himalayas and Adjoint Country. *Indian Forester 60*(10), 1970.P. 710-713.

37. Sherwill, J.L., Journey of a trip undertaken to explore the glaciers of the Khanchanjungha group K in the Sikkim Himalaya in Novermber 1891. *Journal of Asiatic Society Bengal* Vol.31(5), 1892.P.457-479.

38. Sahu, R., *The Himalayan Miscellany,* NBU, West Bengal, Vol 7 &* Mar 1993-94, Vol 11 Mar 1997.

39. *World Health Forum*, WHO Geneva, Vol.14, 1993, Vol.15, 1994.

GOVERNMENT AND ORGANISATIONAL REPORTS

1. Directorate of Census Operation, Sikkim, *District Census Handbook, North, East, South & West Districts,* Series-22, Part XII-A&B, Village and Town Directory, Village and Town wise Primary Census Abstract, 1991.
2. Government of India-Planning Commission, *Report of the Task Force for the Study of Eco-Development in the Himalayan Region,* New Delhi, INSDOC, 1982.
3. Government of Sikkim, Wildlife Circle, Forest Department, *Khangchandzonga (High Altitude) National Park, 1998.*
4. Government of Sikkim, Planning and Development Department, *Report on Tourists in Sikkim,* Bureau of Economics and Statistics, Gangtok, 1992.
5. Government of Sikkim, Planning and Development Department, *Report on Economic Census, 1990, Vol.1.*
6. Government of Sikkim, Department of Health & F.W., *Health Information Bulletin,* Sikkim, 1998.
7. Government of Sikkim, Planning and Development Department, Sikkim, *A Statistical Profile (1979-80 and 1991-92),* Bureau of Economics & Statistics, Gangtok.
8. Government of Sikkim, Sikkim Democratic Front Government, *Sikkim Today.,* 1998.
9. Government of Sikkim, Planning and Development Department, *Report on Assessment of Slums, (1993-94),* Bureau of Economics and Statistics, Gangtok, 1994.
10. India Census 1931, Bengal and Sikkim. Vol 5.
11. Indian Census 1951, West Bengal , Sikkim and Chandannagar Vol. 6.
12. India Census 1961, West Bengal and Sikkim, Part I.
13. India, Census 1961, West Bengal, and Sikkim. Part 2B(II), General Economic Tables.
14. India Census 1971, Sikkim District Census Handbook,
15. India Census 1971, Sikkim Par 2(A) to (F), General population Tables, General Economic Tables, Social and Cultural Tables, Migration Tables and Fertility Tables.
16. India, Census 1991 Sikkim Paper 2 of 1992 Brief Analysis of Primary Census Abstract.
17. India, Census 1991 Sikkim Series 1, paper 1 of 1992(2), Final Population Totals.
18. India, Census 1991 Sikkim Series 1 paper of 1992(I) Final Population Totals.
19. Sikkim Planning and Development Department, Ninth Five year Plan 1997-2002. Draft Proposal from Government of Sikkim.

Index